REPORT

OF

THE SELECT COMMITTEE

OF

THE SENATE OF THE UNITED STATES

ON THE

SICKNESS AND MORTALITY ON BOARD EMIGRANT SHIPS.

AUGUST 2, 1854.

WASHINGTON:
BEVERLEY TUCKER, SENATE PRINTER.
1854.

33d Congress, 1st Session. | [SENATE.] | Rep. Com. No. 386.

IN THE SENATE OF THE UNITED STATES.

August 2, 1854.—Ordered to be printed, and that 5,000 extra copies be printed for the use of the Senate.

Mr. Fish made the following

REPORT.

[To accompany Bill S. 489.]

The Select Committee appointed under an order of the Senate of the United States, passed on the 7th of December, 1853, "*to consider the causes and the extent of the sickness and mortality prevailing on board the emigrant ships on the voyage to this country, and whether any, and what, legislation is needed for the better protection of the health and lives of passengers on board such vessels*"—*to whom were also referred a petition of the American Medical Association, and a petition of inhabitants of New York—respectfully report:*

Under a full sense of the importance of the matter submitted to their consideration, as well as with reference to the business interests of the country as to the cause of humanity at large, the committee have spared no trouble in collecting information from such sources as appeared to them best entitled to their confidence and respect. With a view to the procuration of accurate *data* upon which to form a correct judgment in the premises, a circular was prepared containing fourteen questions relative to the police and sanitary management of emigrant ships, to which specific answers, based as far as practicable upon experience, were requested. Copies of this circular were addressed and sent to members of the medical profession, (whose opportunities of forming correct opinions on the subject had been furnished by experience, and therefore are entitled to consideration,) merchants, navigators who had been engaged in the passenger trade, persons connected with the collection of the customs, presidents of benevolent societies for the relief of indigent emigrants, and others throughout the country and in the cities at which the landing of foreign passengers has chiefly taken place. The answers which have been received to these communications are very numerous, and, in many instances, drawn up with great ability, particularly those from physicians, who, referring to the diseases which have caused the suffering on board of passenger ships, have very properly deemed it necessary to treat of their nature and the causes which have given rise to them.

In disposing of the information thus brought under their consideration, the committee have deemed it best to adopt the course indicated by the order under which they were appointed, and to treat the subject under three distinct heads, to wit:

1st. The diseases on board of emigrant ships, and their causes.

2d. The extent of the sickness and mortality.

3d. The means of prevention.

According to the information laid before the committee, the three diseases by which passenger ships have been chiefly scourged, are—typhus, or ship fever, as it is called when it takes place at sea, cholera, and small-pox.

Of the first named of these, typhus or ship fever, it is said that "it is the product of a miasm as distinct as that of marshes, which produces intermittent fever; and this miasm is itself as necessary a result of certain prior circumstances as the marsh miasm is the product of marshes. And further, the means for its prevention are as clear and controllable in the one case as the other. Thus, if an offensive marsh be thoroughly drained and dried, its peculiar miasm, and the disease which it caused, will disappear; and so, by preventing the formation of the miasm of ship fever, (as easy of accomplishment as the other,) that disease will in like manner be prevented or avoided."

The circumstances referred to are stated to be:

1st. The confinement of a number of people together in apartments disproportioned in size to their requirements of wholesome respiration.

2d. The retention in the same apartments of the excretions from the bodies of individuals thus confined; such as the matter of perspiration, the carbonic acid gas and moisture from the breath, and other more offensive excretions. These, acted on by the artificial heat of the apartment, or even by the natural heat of the bodies alone, will become decomposed and produce an effluvium which will react poisonously on the persons living in it.

3d. Too great exclusion of pure air.

Some, if not all, of these causes are found to exist on board of every passenger ship in a greater or less degree, and the consequence has been the mortality which has taken place of late years, the weakness and enfeebled condition of emigrants generally at the time of embarkation, operating as a predisposing cause of the evil, by rendering them less able to resist the impression of the poison existing on board of ship.

It appears to be a peculiarity of this miasm that it attaches itself to everything that it touches—to clothing, bedding, furniture, and the walls of apartments—by which it is absorbed, and becomes more virulent in its action in proportion to the length of time during which it is permitted to remain. It is stated, upon the highest authority, that this poison may last in "fomites" for six months, and even for two or three years. Dr. Wilson Phillip, in his standard and elaborate work on fevers, in his chapter on typhus, says:

"Fomites often retain contagion for a great length of time, and may convey it any distance. It is a general opinion that *fomites* more readily communicate the disease, and convey it in a *worse form*, than the sick themselves." Dr. Cullen says "the effluvium constantly arising from the living human body, if long retained in the same place without being diffused in the atmosphere, acquires a singular virulence." And Dr. Parr, in his medical dictionary, asserts that "fevers caught by recent infection are mild compared with those which arise from contagion long pent up, styled *fomites*."

In his examination before a committee of the British House of Commons, Dr. William Gladstone states that "men-of-war were formerly ballasted with shingles; that this ballast was often not shifted for many years; that when it was turned out it produced fever in several of the ships; and that this fever assumed the character of the prevailing fever of the station, (whatever it might be,) at which the vessel happened to be at the time."

Cholera.—The second named of the diseases by which passenger ships have been infested, is of Asiatic origin. It first made its appearance in western Europe, in or about the year 1832, and committed its fearful ravages without regard to the rank or station of its victims. What at first appeared to be but temporary, has, of later years, assumed the character of a regular disease, to which mankind are subject, in common with the other maladies that flesh is heir to. In one respect, it seems to differ from the typhus fever, being generally *an epidemic, which typhus never is.* The open air always puts an end to typhus or ship fever, whereas, cholera is controlled by no such corrective, but wings its deadly flight over the prairie and the prison-house alike. Although this fearful disorder confines itself to no precise localities, there appear to be circumstances under which it is peculiarly apt to make its appearance. These circumstances have been ascertained to be in a great degree similar to those which give rise to typhus fever. The poor and vicious, whose vital powers are enfeebled by want of wholesome nutritious food and close confinement, or criminal excess, are found to be much more liable to its fury than persons who have good nourishing food in abundance, take regular exercise, and abstain from indulgences that weaken the general tone of the system, whilst they add to the nervous excitability of the body. Dr. Griscom, of New York, whose opportunities of observation have been very numerous as agent of the board of commissioners of emigration, to whom the charge of this class of patients is entrusted, thinks that the miasm of typhus is the *direct* product of the vitiated excretions of the human body, pent up within a small space, and made to engender a malaria, the inhalation of which, to a certain degree, produces this peculiar disease; whereas cholera, by disabling the system, renders it liable to be overcome by the choleric poison at an earlier period than that at which typhus fever makes its appearance. Your committee do not propose to discuss this point, as it is not necessary to inquire whether human life be destroyed directly or indirectly, provided that exemption from destruction can be procured by the removal of the exciting causes of disease and death. Cholera, it is true, often appears and disappears without any apparent cause, a fact the reason for which is still hidden from the eye of science, and can only be explained by time and experience. It is sufficient to know that if the body be kept in a healthy well-balanced condition, and its functions be not interrupted by any disturbing causes, it may, in the generality of cases, bid defiance to the assaults of disease, in these or any other forms. It is believed that the same rules of conduct are necessary for the prevention, so far as practicable, of both of these maladies, with perhaps the exception of those relating to disinfection; unless, as some suppose, cholera be contagious, as well as typhus fever. As to the theories, entertained by some, that cholera on ship board arises from the virus of the disease having been imbibed by the persons or clothing of passengers, previously to embarcation, or that it is met with in certain zones, through which the ships pass in reaching the western continent, your committee will only remark that all that can be done by the owners of passenger ships is to prevent the existence of any exciting cause of sickness on board of them, or of any state of things by which it may be nourished and sustained if contracted elsewhere. In reference to the idea that disease is caused by passing through certain zones, it is proper to remark, that the hypothesis appears to be contradicted by the fact, that it frequently happens that vessels leaving the same port on or about the same day, and arriving at their point of destination about the same time, are differently affected by sickness. If there be anything in the atmosphere of particular zones or belts, it must be encountered alike by ships sailing probably within a few miles of each other, propelled by the same winds and standing on the same courses. Admitting the poison to be in the air, the natural inference would be that the same cause would affect all ships within the same limits, similarly situated on the bosom of the ocean, in the same way; and that sickness would be produced on board of all alike. Such, however, appears not to be the case. The ship Lucy Thompson, after a passage of 29 days, arrived at New York, from Liverpool, on the 11th of September, 1853, with the loss of 40 out of 835 passengers by cholera. The William Stetson, arrived on the same day, after a passage of 31 days, with 355 passengers, having lost none on the passage, and the Great Western arrived on the day previous, 10th September, after a passage of 31 days, with 832 passengers, no death having occurred on board. On the 19th of September, 1853, the Isaac Webb arrived at New York from Liverpool, with 773 passengers, 77 having died of cholera, after a passage of 29 days. On the next day, the Roscius arrived from the same port with 495 passengers, after a passage of 35 days, 6 days longer than that of the Isaac Webb, no death having taken place. On the 21st of September the Leviathan arrived at New York, from the same port, after a passage of 35 days, with 559 passengers, and without any deaths on board, and the Northern Chief, arrived on the same day from the same port, after a passage of the same, with 626 passengers, and no deaths.

On the 15th of October the Montezuma arrived at New York, from Liverpool, in forty-one days, with four hundred and four passengers, and a loss of two; while the Marmion arrived on the same day, after a passage of twenty-five days, with two hundred and ninety-five passengers, and a loss of thirty-six by cholera.

On the 29th of October the Lady Franklin arrived, after a passage of forty-four days, with seven hundred and forty-one passengers, and without loss; while the Albert Gallatin, which sailed from the same port on the same day with the Lady Franklin, arrived on the 30th, the day following, after a passage of forty-five days, with seven hundred and fifty-six passengers, lost thirty-eight by cholera.

On the 1st of November the Constitution arrived at New York, from Liverpool, after a passage of fifty-three days, with six hundred and seventy-one passengers, and without loss; while the Forest King arrived on the same day, after a passage of forty-eight days, with five hundred and fifty-eight passengers, and a loss of forty-two by cholera.

The New York arrived at New York, from Liverpool, on the 21st of October, after a passage of forty-three days, with a loss of sixteen out of four hundred and thirty-four passengers. The Progress arrived on the same day, from the same port, after a passage of forty-five days, and with a loss of seventeen out of four hundred and twenty-eight passengers by cholera; while the William Nelson, which arrived on the same day, after a passage of forty-six days, with four hundred and twenty-eight passengers, had not a single death. The State Rights also arrived on the same day, after forty-seven days' passage, with three hundred and sixty-two passengers, and without a death.

The Washington arrived at New York on the 23d of October, after a passage of forty-one days, with nine hundred and fifty-two passengers, and a loss of eighty-one; while the Guy Mannering arrived on the 25th of the same month, after thirty-seven days' passage, with seven hundred and eighty-one passengers, and without loss.

These examples might be multiplied almost at pleasure, showing that vessels which left the same port almost at the same time, and reached the same point of destination about the same time, and consequently may be supposed to have been in the same latitudes, and subject to the same winds at the same moment, suffered very differently. If the state of the external air produced the disease, why, your committee would ask, were not ships similarly situated affected in the same manner, and to something like the same extent? The cases presented show to the minds of the committee conclusively that the disease on board of these vessels is to be attributed to some exciting cause which existed within themselves, and cannot be referred, with any show of reason, to the condition of the atmosphere or the prevalence of certain winds on the ocean.

Of small pox—the last named of the three diseases from which passenger ships have suffered most extensively—no description is deemed necessary, as, unfortunately, its peculiar characteristics are but too well known. The frightful ravages of this disorder, prior to the discovery by the great benefactor to his kind, (Jenner,) that the introduction of the vaccine virus would produce a disease which would disarm small pox of more than half of its terrors, have left traces in the Old World not to be effaced and never to be forgotten. It is sufficient to say that the rules which apply to the prevention of typhus or ship fever and cholera are, in the main, also applicable in the case of small pox.

The next branch of the subject to which the committee would respectfully call the attention of the Senate is the extent of the sickness and mortality on board of emigrant ships.

In prosecuting their inquiries and investigations upon this head, the committee have been met by many and serious difficulties. The sources of information to which they looked with most confidence for accurate details have been the records of the custom-houses at the ports at which the greater portion of the emigrant ships arrive. Even in that quarter disappointment has to an important extent attended their labors, although circulars were addressed to the collectors of the ports of Boston, New York, Philadelphia, Baltimore, Charleston, and New Orleans, and also to the municipal and local authorities at some of these and other places and elsewhere. The gentlemen to whom these circulars were addressed have responded to the call thus made upon them with a promptness and alacrity which do them honor, and lay the committee under great obligation. Unfortunately, however, the returns received are not so full, detailed, and accurate as could have been desired. This deficiency is not to be referred to any want of care on the part of the officers alluded to, whose reports are in accordance with the requisitions of the acts of Congress in which their respective duties are defined, but rather to the peculiar character of the information required by the committee, and to the insufficiency and inaccuracy of the reports made to the collectors by the masters of vessels. The requisitions under the laws for the collection of the revenue having for their object, in the first place, the ascertainment of the value of the merchandise to be assessed with a view to the revenue to be collected, the returns relating to the health of passengers have engaged but a secondary importance.

Generally speaking, the health of those "who go down to the sea in ships" has been so great, until of late years, that the mortality prevailing on ship board has not excited interest enough to require sufficient statistics in regard to it to form a part of the regular returns to the government. Hence it has occurred that the reports at the various custom-houses have not been either accurately or uniformly made. Even at New York, the great commercial emporium of the country, the list of arrivals furnished by the collector does not always state the number of deaths, while it omits entirely to notice those who are landed sick, a point of much importance in connection with the duties of the committee. In this respect the return of the health officer of New York, who holds his appointment under the authority of that State, is more full and satisfactory. In order to overcome the difficulty that here presented itself, the committee determined to adopt as their guide the report of the collector, who, being an officer of the federal government, may be considered as more immediately connected with Congress, and at the same time to note the discrepancies between it and the report of the health officer, adopting such features of the latter as are not presented in the former.

With this view a table has been prepared, marked C, including the last four months of the year 1853, during which the sickness and mortality prevailed to the greatest extent, shewing the number of arrivals at New York of passenger ships, the number of passengers according to the report of the collector, the number of deaths during the passage as stated by him, designating the number reported to have arisen from cholera, and the number landed sick at New York, as stated by the health officer, so as to present at one glance the state of the whole matter.

It will be seen by table marked A that, according to the annual reports of the Secretary of State to Congress, the whole number of passengers arrived in the United States during the last ten years, say from October 1, 1843, to December 31, 1853, is 2,270,847. These statements do not give the number of deaths during the passage, nor of those landed sick. According to the above authority the number of arrivals in 1853 was 400,777.

The reports of the collectors of the ports of Boston, New York, Philadelphia, Baltimore, Charleston, and New Orleans show an aggregate of passengers arrived at those ports during the year 1853 of 370,902, and of deaths during passage of 2,444, as will be seen on reference to table marked —, the returns made by the collectors to the Secretary of State under the act of 1819, and embodied in his annual report to Congress, (Ex. Doc. No. 78, H. R., 1st sess. 33d Cong.,) show the number of arrivals during the year at those ports to have been 395,325. In the reports from Boston and New York the aggregates, as given by the superintendent of alien passengers at the former, and the health officer at the latter, do not agree with those of the collectors, but, for the reasons already assigned, the committee have thought proper to base their calculations upon the returns of the collectors, they being officers of the federal government.

From the above returns it appears that the number of deaths, as well as the per centage of mortality among the passengers on board of New York vessels, has been considerably greater than those of vessels trading to Boston and other ports. This might have been, perhaps, expected, for a variety of reasons. New York being the great commercial emporium of the Union, passengers from every country in Europe have been induced to regard it as the point to which they should direct their courses. Hence the huge structures furnished by the enterprise of that great metropolis for the transportation of passengers have been crowded to excess, and, as a necessary consequence, the causes of disease have existed on board of those vessels to a greater extent than on any other. In general the per centage of deaths decreases in proportion as the numbers of passengers are less, and it is found that where passengers have been distributed in smaller numbers, disease and death have been less prevalent. The table shows that the smallest per centage of deaths has occurred on vessels from ports of Europe other than those of Liverpool, London, Bremen, Hamburgh, and Havre, which, being out of the great thoroughfares of commerce, have presented fewer attractions to the great mass of emigrants. The vessels from these ports, being less crowded, are more easily ventilated and kept clean, and present greater facilities for the proper preparation of the food of the passengers, and for their exercise in good weather.

During the four last months of 1853, 312 vessels arrived at New York from European ports, with 96,950 passengers. Of these vessels 47 were visited by cholera, and 1,933 passengers died at sea, while 457 were sent to the hospitals on landing, there, in all probability, to terminate in a short time their miserable existence, making nearly two per cent. of deaths among the whole number of persons who had embarked for the New World, and nearly two and a half per cent. if those who were landed sick be included. On board of the 47 vessels attacked by cholera the number of passengers was 21,857, of whom 1,821 (being 8.48 per cent.) died on the passage, and 284 were landed sick, making 9.68 per cent. of dead and diseased in an average passage of 39 days. (See annexed table marked C.)

Of the arrivals above mentioned 112 were from Liverpool, with an average of 435 passengers on each. Twenty-four of these vessels, with an average of 577 passengers, or an average excess of 142 passengers each over the general average of the whole number of vessels, had cholera on board.

Of twenty vessels which arrived from London, five had cholera on board. The average number of passengers on board the vessels attacked by cholera was 411 each, while that of the whole number was but 326.

Of fifty-two vessels which arrived from Bremen, three had cholera on board. The average of passengers each on board of the whole number of vessels was 201, while the average on board of those attacked by cholera was 259.

Of twenty-two vessels that arrived from Hamburg, six had cholera.

Of forty-two vessels which arrived from Havre, six had cholera. The average on board of the whole number of ships was 409, while on board of the six cholera ships the average was 561.

Of sixty-four vessels which arrived from other ports of Europe, three had cholera on board. The average of passengers on board of the whole number of these ships was 148, while that on board of the ships attacked by cholera was 185.

The average on board of the whole number of vessels (312) that arrived from Europe during the four months was 311, while the average on board of the forty-seven that had cholera was 465. The average on board of the vessels which arrived, exclusive of those with cholera, was 283, showing that the cholera vessels carried an average excess of 182 each over those that were in comparison healthy.

Of the vessels which escaped from cholera, there were thirty-three—carrying on an average 335 passengers each—on board of which deaths occurred. On these vessels, the number of deaths was 112 out of 11,044 passengers.

It appears from the above statement of facts, that the ships on board of which cholera broke out were those which were most crowded with passengers, and that the vessels on board of which deaths from other diseases occurred were the next most crowded, whilst the residue, which were healthy, had the lowest average of passengers.

These facts speak for themselves, and make any commentary on the part of the committee unnecessary. In connexion with the above, it may be stated that nearly twenty-one and one-half per cent. of the ships from Liverpool were attacked by cholera; twenty-five per cent. of those from London; less than six per cent. of those from Bremen; nearly thirty-three and one-third per cent. of those from Hamburgh; upwards of twelve per cent. of those from Havre; and only about four and one-half per cent. of those from other ports of Europe were visited by it. From this, it would seem that this frightful malady delights in pursuing the great thoroughfares of commerce and international intercourse, while, comparatively speaking, it overlooks the bye paths along which mankind pursue their way in smaller groups.

While medical men all agree that the want of fresh wholesome air on board of passenger ships is one of the most prominent causes, if not the most so, of the sad mortality which has prevailed, there seems to be much diversity of opinion as to the sufficiency of the means of ventilation as prescribed by the laws in existence, as well as the space to be allotted to each adult passenger. The space at present required by the laws of the United States, is fourteen superficial feet, when the height, between decks, is not less than six feet; sixteen feet when it is less than six feet and not less than five feet; or twenty-two feet when it is less than five feet. As it appears to your committee, the quantity of wholesome respirable air to be secured to the occupant or occupants of the apartment assigned to them is the important point to be settled, space should be calculated with reference to ventilation. A room of ten feet, thoroughly well ventilated, may be occupied with health and comfort by a given number, when a room of twice the size, shut up and deprived of a current of air, will become unwholesome in the course of a few hours, from the want of oxygen, and be injurious to the health of the same amount of occupants. The atmosphere, when charged with the effluvia caused by the excretions of the human body, if inhaled into the lungs, will inevitably produce disease and death. So, with the same means of ventilation, apartments situated in different parts of a ship, may, or rather must be unequally ventilated, and some of them may be unhealthy while others are perfectly wholesome. From this it is evident that no general rule can be equally effective, and reach every case, inasmuch as the peculiar conformation and internal arrangements of each ship must have a very material influence in ascertaining, with exactness, the extent of the means to be used in effecting a thorough ventilation. The opinions furnished to the committee, in regard to space, are in a high degree conflicting, some of them being in favor of the sufficiency of the provisions contained in the present laws, while others pronounce them to be totally inadequate. This antagonism of opinion may, in part at least, be accounted for by the fact that the interests of those from whom they emanate are opposed. Generally speaking, the parties to be accommodated and their friends deem the present provisions insufficient, while those who are required to incur the expenses of a change, with a few exceptions, think differently. For instance, some of those who have been consulted say that the present tubular ventilators answer the purpose for which they are intended, while others insist that they are totally inadequate, and that windsails should be used; while others again assert that the noxious gasses enter into the various openings and corners of the apartments, and can only be gotten rid of by the use of a powerful forcing apparatus. Of windsails, some again allege that, in practice, they are useless, as those of the passengers who happen to come under the influence of the current of air are rendered uncomfortable by it and tie them up, and thus render them totally inoperative. On the respective merits of these modes of ventilation the committee are not sufficiently informed to express a judgment.

The committee would, however, state that it appears to them reasonable to suppose that the air which is loaded with the exhalations from the human body finds its way into the holes and recesses formed by the permanent structure of the ship, the berths, (placed, as they generally are, transversely, and having small spaces underneath them,) and the luggage of the passengers. These recesses cannot, it would seem, be materially affected by any ordinary current of air passing gently through the apartment, and can only be reached and cleansed of their contents by the operation of some appliance which shall cause the volume of fresh air to be distributed to every portion of the apartment occupied by the passengers. The committee would further remark, in connexion with this part of the subject, that the transverse position of the berths alluded to, while, so far as they can see, it can answer no good end, must contribute very much to increase the number of places inaccessible to ordinary means of ventilation, and thereby tend to render the sleeping apartments of passengers unwholesome. It would further seem, that the receiving and discharging ends of the tubular ventilators, being at about the same elevation, the current of air through them must be of very moderate velocity, and, notwithstanding the progress of the vessel, which during light winds is slow, scarcely sufficient to expel from the nooks and corners of the hold the noxious gasses, the inhalation of which produces disease. Under such circumstances, the committee repeat the opinion, that nothing short of some powerful agency will be effective in expelling the foul air, and enabling passengers to inhale a wholesome and healthful atmosphere. In connexion with the subject of ventilation, the committee would respectfully ask the attention of the Senate to the occupation of the orlop or lowest deck in ships having three decks, upon which emigrants are frequently stowed in passenger vessels. Notwithstanding the almost total absence of fresh air or light in these apartments, the desire to derive profit from every portion of the vessel has appropriated even this to the occupation of passengers. According to the statements of persons well informed on the subject, these places, when occupied by passengers, not only become the depositories of the most noisome filth, which can neither be reached nor removed, but they give rise to a stench and effluvium, which, rising to the upper decks, tend to render them unhealthy. Some of the persons from whom communications have been received, are of opinion that the orlop deck may be appropriated to passengers with safety, provided the space allotted to each be considerably increased; but the almost universal opinion appears to be, that the use of it for this purpose should be prohibited by law. When the form of the immense structures built for passenger ships is considered, it is evident that the apartment alluded to must be almost entirely deprived of air and light, as it lies next to the kelson, and is entirely under water, and it appears almost incomprehensible that human beings should be required to occupy it. The committee have no hesitation in coinciding in opinion with those who recommend the prohibition of the use of the orlop deck as a sleeping apartment, and have, therefore, inserted in the law herewith reported a provision to that effect.

With reference to the supply of wholesome air on board of passenger ships, the following remarks of Dr. Griscom, of New York, are so judicious and practical, and occuy so small a space that the committee do not hesitate to insert them entire.

"The next point of inquiry will be the extent of cubic area, which each passenger should be allowed. Several elements are found pertaining to this calculation. On the principle before laid down, 'that the more active the ventilation of an apartment, the more it may be crowded with impunity,' it will at once be seen that in the allotment of space for passengers, the height of, and the freedom of circulation of air through the steerage should be taken into account. The application of this principle would have a salutary effect upon the owners of passenger ships, in inducing them to make the steerages of the greatest possible height, and introduce the most effective means of ventilation, as thereby they would obtain the privilege of the greater number of passengers.

"But as in the case of the height of the ceiling, a minimum amount of free cubic space should be affixed by law. In relation to the question of what this should be, a diversity of opinion exists among well informed men. It involves the point of the amount of air requisite for the wholesome respiration of an individual for a given space of time. The estimates of the amount required for an adult *per minute*, and which at the end of that time must be entirely removed to avoid the risk of reinhalation, vary from four to ten cubic feet. This with the air at rest and not communicating with the general atmosphere. With a free communication between the general atmosphere and that of the apartment, whereby the carbonic acid gas and other exhalations can be freely diffused abroad, ten cubic feet per minute is probably more than is needed, but the lower figure given above is, on the other hand, too small. The famous Black Hole of Calcutta, we are informed, was about eighteen feet square, and it was probably not over ten feet high. This gives it an area of 3,240 cubic feet. On the fatal night which has given it so great notoriety, it was made to contain 146 persons, thus allowing to each only 22 cubic feet, though as each body excelled an equal bulk of air, it was probably not near so much as that. At the lowest estimate mentioned above, (4 feet,) before five minutes elapsed after the door was shut upon them, the hapless victims began to reinhale their own exhalations, and this process was, of course, repeated, at each similar successive period, until death began to reduce their numbers. They were confined there about ten hours, and although there was an open window on one side, twenty-three only survived till morning, and they were in a "high putrid fever," (typhus.)

"I allude to this oft quoted case, chiefly for the purpose of showing those not familiar with this topic, and who may be sufficiently interested in this subject to read this communication, what are the true principles upon which a calculation for the allotment of space for emigrant passengers should be based. Any more minute detail in this already too long paper of the elements of this calculation, would probably be considered as burdensome. I may state, in concluding this part of the subject, that a recent examination of the two steerages of one of the largest packets belonging to this port, New York, (authorized by the present law to carry over nine hundred,) gave as the cubic feet for each passenger, not deducting the room occupied by the necessary solid contents of the bodies of the passengers, for the upper apartment 103 feet, and for the lower 112 feet. This vessel, on her last homeward voyage, lost one hundred passengers at sea.

"In my opinion, not less than 250 to 300 cubic feet should be given to each passenger."

With reference to the cooking arrangements, the information furnished to the committee is contradictory, but the weight of opinion is clearly in favor of the food of the passengers being not only furnished, but also cooked by persons belonging to the vessel. Independently of the fact, that the food can be better done and more regularly served by an agent of the ship than by persons entirely ignorant of cookery, and suffering under sickness and other ailments, not to mention the depression of mind attendant on leaving the homes of their nativity forever, this arrangement has the advantage of placing in the hands of the captain a very effective method of enforcing good order and cleanliness among the passengers. To cook for hundreds of persons, even with the most ample accommodations, requires regularity and system, but when every man becomes his own cook, and prepares his food whenever whim or inclination may dictate, the proper preparation of the provisions is next to an impossibility. The apparatus for cooking on board of ships of the largest size, consists of a caboose in the proportion of four feet long and one and a half feet wide for every two hundred passengers. At this caboose all the victuals must be cooked between certain hours, as the fire is put out at a fixed time to ensure the safety of the vessel. In attempting to effect this, the sick are brought into conflict with the healthy, and the weak with the strong, and when such is the case, it is an easy matter to imagine who are the sufferers. The sick, to whom nutricious food properly prepared is essential, have no chance whatever in such a contest, and are forced to retire without cooking their food at all, or after preparing it only in such a manner as to make it the cause of sickness, and, perhaps, death. Even with the robust and strong the use of half cooked food is almost sure to produce indisposition; but when invalids, laboring under affections of the stomach and intestines, are obliged to take it, the inevitable consequence is serious if not fatal disease. So far as cholera is concerned, experience has shown that nothing will produce it, in cases of predisposition that way, sooner than the consumption of meats or vegetables improperly cooked. What chance then, the committee would ask, have these miserable creatures, closely confined, and breathing a noisome atmosphere, sea-sick, and depressed in spirit, and withal required to prepare their own food, from which they are prevented by the selfish and hard-hearted.

In view of these facts the committee are of opinion not only that the provisions should be furnished by the ship, as when purchased in large quantities by competent judges, they are more likely to be of a good and uniform quality; but the cooking should also be done by the ship's cook or his assistants, and the food served out at certain hours of the day to passengers individually, or to messes, as may be found most convenient. In case messes are formed, the distribution can be made to caterers appointed by each, who can, in turn, divide the food among the members of their respective messes. Such an arrangement would possess the advantage of uniformity and regularity, and prevent all impositions by the strong and healthy upon the feeble and infirm, and, as a consequence, be the means of avoiding the heart-burnings and ill-feeling which cannot fail to attend a competition for the use of the fire.

With regard to the rule regulating the distribution of passengers, or rather the apportionment of them, two modes present themselves by which the end may be effected. One of these modes is to graduate the number to be received by the tonnage of the vessel, by allowing a passenger for every given number of tons; the other is to allow to each passenger a certain number of superficial feet, reference being had to the height of the apartments. The first of these standards was formerly in use, but was abandoned on account of the inequality of register tonnage, growing out of the difference of the models according to which ships are constructed. It is universally known that the vessels of this country are modelled very differently from those of Europe generally, and more especially those of Great Britain. American ships being built for speed, are usually much sharper than those of the United Kingdom, which are constructed more with a view to capacity for cargo than making quick passages. This difference is said to have arisen, in a measure, from a desire on the part of British ship owners to avoid the tonnage duty, which is graduated according to measurements which do not involve the *actual capacity* for cargo; a vessel of a given register tonnage, being capable of receiving many more tons of cargo than the register would indicate. Hence it is evident that a ship built after the British fashion would have much more space for passengers between decks than one constructed after the American style, and with a view to making quick passages; the former being full and rounded, while the latter is sharp, and if the expression may be allowed, *lean* in its figure, and of a much greater draught of water in proportion to its length and breadth of beam. For a practical illustration of what is here stated it is only necessary to compare the American clipper-built ships, now so much in vogue, with the heavy, capacious vessels in use in the British commercial marine. To avoid this inequality of operation involved in the law as it formerly stood, the standard of allowance was changed so as to throw tonnage out of the calculation, and to allot to each adult passenger fourteen superficial feet of deck, in cases where the height between decks is not less than six feet; sixteen feet when less than six, but more than five in height, and twenty-two feet when less than five feet in height. Of the expediency of the above change, serious doubts have been entertained by very sensible practical men, as in many instances it has been found mischievous, by allowing ships to carry enormous crowds of passengers, for whom there was no accommodation consistent with their comfort or health. Experience seems to indicate a combination of the two modes as the best course, as by adopting the tonnage feature, the over-crowding of ships will be prevented, while on the other hand, the ascertainment of the exact space to be allotted to each passenger will tend much to the promotion of his comfort and the preservation of his health.

The information laid before the committee in regard to the propriety of employing regular physicians, and nurses, and hospital assistants on board of passenger ships, goes to confirm the impressions already entertained by them upon that point. There are very many reasons why such an arrangement should exist. Such an officer would be of incalculable service, not only in ascertaining the condition of the vessel, with regard to the health of the passengers, but in prescribing the best mode of preventing or overcoming evils connected with the sanitary administration which none but an experienced eye can detect. The statements of such an officer, in regard to deaths happening on board, would, of course, be of great weight, and frequently prevent erroneous impressions, affecting the commander of the vessel and his officers. The observations of an experienced physician, in cases of visitations, by disease, such as have occurred within the last year or two, would be of great value to the world of medical science, and aid in an essential degree, the cause of philosophy and general knowledge. For these, and other reasons, the committee are decidedly of opinion that such an arrangement would be desirable; but taking into view the uncertainty of the number of passengers that may be on board, and other circumstances, they cannot see how it can properly be made the subject of effective legislation.

As to the propriety of separating the passengers, according to sex, so far as may be consistent with the union of families, with a view to the prevention of the shocking immoralities said to be practised on board of passenger ships, it is believed there can be but one opinion; but unfortunately, the committee have, in the course of their investigations, been forced to entertain great doubts of the practicability of such a measure, owing, in part, to the existing mode of constructing vessels. The entire separation proposed can only be effected, as the committee think, by the construction of permanent partitions which shall divide from each other the apartments allotted to the sexes respectively, and the establishment of separate entrances to these apartments. Now, it is acknowledged on all hands, that proper ventilation is indispensable to the well being and health of persons crowded together as these people are on board of these ships. In order to ventilate these vessels properly throughout the steerage, with the means now used, it is necessary that the current of fresh air shall be made to pass from one end of the ship to the other. The permanent partitions intended to divide the apartments of the different sexes must, of necessity, be constructed transversely, and cannot fail to prevent the passage of air above alluded to; a difficulty, which, so far as the committee are aware, can only be obviated by the erection of slat partitions, which would impair the privacy which is so much desired, at the same time that it would interpose some obstruction to the free current of air through the steerage.

In the second place, with respect to the separate entrances to the different apartments, it is believed that vessels are uniformly built with but two hatchways, and that to effect the object proposed, it will be necessary to introduce a third for the especial purpose. This might perhaps be done in vessels hereafter built, but as any change in the law must operate on the vessels now in existence, the question presents itself—what is to be done with the ships now in the passenger trade? To this question the committee can find no satisfactory answer, and therefore would express their fears that, however desirable, the measure is difficult of attainment by specific legislation. In coming to this conclusion, the committee feel great regret, as they, in common with all well-thinking persons, view with a feeling amounting to disgust and horror, the improper intercourse said frequently to exist, not only between passengers of different sexes, but between the crew and female passengers, whose peculiar situation renders them accessible to the advances of the dissolute and unprincipled.

The committee are, however, happy to believe that the subject is in a great degree within the power of control by the master of the ship, and that an assignment of births in different parts of the vessel can be made, separated so as not to interfere with a sufficient ventilation, but protected against intrusion by a well considered and well enforced system of police on board ship. The details of such separation and of such system of police cannot be prescribed by legislation, but must be left for the present to the captain or master of the ship, on whom devolves a weight of moral responsibility in case of neglect, far greater than the responsibility arising from the property entrusted to his charge.

If this intercourse can be prevented, or even abated, it becomes the absolute duty of every commander of a ship to institute rules to that effect, which shall be enforced with the utmost strictness; and a failure on his part to do so should be regarded by his employer as a sufficient cause for severe censure, if not dismissal. Neglect or omission in this respect involves the ship owner in just censure, as a watchfulness on his part in regard to the adoption of proper regulations, and of a proper police on board of his ship, would do much towards effecting an immediate cure of the evil; while the committee believe, in the construction of ships hereafter to be built, an effectual separation may be ensured, while a thorough ventilation may at the same time be secured for each of the separate apartments.

Having thus disposed of the topics presented to their consideration by the answers to the questions contained in their circular, the committee will now ask attention to a suggestion which had previously engaged their serious consideration, but was forcibly presented to their notice by the Board of Health of Philadelphia, and which, in their opinion, is entitled to the greatest and most favorable attention. The great difficulty of preparing a general law which shall embrace in its operation all of the minute points involved in the management of passenger ships, and effectively place the passenger and ship-owner on the best possible footing, must present itself to every mind. In the first place, our country being the general receptacle of emigrants from almost every country of Europe, it becomes necessary so to legislate as to avoid coming into conflict with the legislation of the countries whence these people come, and at the same time protect their interests and the interests of American citizens at home and abroad. Great Britain,* France, the various nations of Germany, the Hanseatic cities, have all of them their passenger laws, in which they prescribe the terms upon those who emigrate from among them shall be conveyed across the ocean. These laws or regulations include not only the equipment and nautical management of the ships engaged in this trade, but also prescribe such dietary provisions as to their makers respectively appear essential to the health and comfort of passengers. To legislate in this country, therefore, so as not to come into conflict with these foreign enactments on some of the many minute points which present themselves in the treatment of such a subject, requires an extent and accuracy of information on details difficult to attain, and would involve a minuteness and variety of legislative enactment suited

---

* The emigrant passenger act of Great Britain was communicated to the Senate of the United States, with a message of the President of the United States. at the present session, and was printed. (See Ex. Doc. No. 58, 1st session, 33d Congress, Senate.)

to the local requirements of each country whence emigrants seek a home on our shores, and liable to become oppressive whenever a change of policy suggests a change in those foreign enactments. It is, of course, the intention of every law-maker that the laws passed by him shall be enforced; and to place enactments on the statute book which cannot be carried into effect, without subjecting those who are governed by them to difficulty and annoyance from foreign countries, is worse than useless. That such a state of things at present exists, under the passenger laws now in force, has been proved by experience; and those engaged in the transportation of passengers do not hesitate to say that the laws under which they are acting are in many respects impracticable, unless at great expense and loss to the American ship-owner. It is only necessary to allude to one case of the kind by way of illustration. The acts of Congress require that every passenger ship shall be furnished with provisions of a certain description to a given amount. An American ship goes to Bremen or Hamburg and there takes on board a cargo of passengers. The laws of those cities require that all passenger ships sailing from them with passengers shall be supplied with a prescribed amount of certain provisions, which are specified. Now, in order to comply with the regulation of the port whence he sails for America, the American captain must provide a supply of the articles required by the local law, and, at the same time, to comply with the law of his own country, he must be provided with the food called for by the act of Congress. This single case involves a double expense in the provisions made for the subsistence of passengers. To obviate this difficulty, it has been the desire of the committee to present a law which shall, in its general operation, provide for the safety and comfort of passengers, and, at the same time, leave the management of minute details in the hands of those whose interest as well as business it is to be thoroughly acquainted with them. It is utterly impossible for Congress to know what is required in each case to make a ship comfortable and healthy. One ship may be ventilated and made perfectly wholesome for passengers by a process which in another vessel, and under different circumstances, would be totally inadequate. It would, therefore, be best, in the opinion of the committee, to leave the means by which ships are made safe and comfortable mainly at the disposal of their owners, and, at the same time, to make them responsible for any untoward results that may attend the administration of their own affairs. It is under this point of view that the suggestion to which reference has been made assumes a degree of importance which entitles it to the greatest and most favorable consideration of Congress.

In a report on quarantine, presented to both houses of Parliament by command of her Majesty the Queen of England, published in London in 1849, signed by Lords Carlisle and Ashby, and Edwin Chadwick and T. Southward Smith, in which the subject is elaborately discussed, the following remarks occur:

"Nevertheless, the experience of the mercantile navy itself affords an example of the successful working of a large preventive principle of jurisprudence; that is to say, *the principle of concentrating responsibility on those who have a direct interest in prevention, and who possess the best means of securing it.* We would request attention to those examples as subjects for international consideration."

The report then goes on to say:

"It is stated that when the system of transportation was first adopted, in some of the earlier voyages, full one-half of those who embarked were lost; later, on the passage to New South Wales, as in the 'Hillsborough,' out of three hundred and six who embarked one hundred were lost; and, in another ship, the 'Atlas,' out of one hundred and seventy-five embarked sixty-one were lost. Yet there were no omissions palpable to common observation, or which could be distinctly proved as matter of crimination to which responsibility might be attached. The shippers were, no doubt, honorable men, chargeable with no conscious designs against the lives of the human beings committed to their care, and with no unusual omissions; but their thoughts were directed by their interests, exclusively to profits; they got as much freight as they could, and they saw no reason why convicts or emigrants should not put up with temporary inconveniences to make room for cargo.

"By a simple change, (based on the principle of self interest, the most uniform, general, and when properly directed, really beneficent of all principles of action,) by the short alteration of the terms of the contract, so as to apply the motive where alone there was the effectual means of prevention, by engaging to pay for those *landed* alive, instead of paying for all those *embarked*, these extreme horrors were arrested; the generation of extensive mortal epidemics was, in a short time, prevented, and clean bills of health might have been given to all the ships which before would have been entitled to none. From the report of the select committee on transportation, in the year 1812, it appears that in one period, namely, from 1795 to 1801, out of three thousand eight hundred and thirty-three convicts embarked three hundred and eighty-five died, being nearly one in ten. But since 1801, after the principle of responsibility began to be applied, out of two thousand three hundred and ninety-eight embarked only fifty-two have died, being one in forty-six. The improvement has continued up to the present time, when it amounts only to one and a-half per cent., or even lower than the average mortality of such a class living on shore. The shippers themselves, without any legislative provisions, or any official supervision or regulations thereto, appointed medical officers or surgeons, and put the whole of the convicts under their charge; the shippers attested their own sense of the propriety, sound policy, and efficiency of the principle, by voluntarily adopting it and applying it to each ship surgeon in charge, whose remuneration was made dependent upon the number of passengers landed alive."

The same principle has been adopted in contracts for the transportation of troops, and found to operate equally well, also to the transportation of pauper emigrants, with a like result; and the authors of the report think the adoption of it will be general in the transportation of passengers, as it puts an end to all the difficulties usually occurring between the passengers and the officers of the ship, who are, of course, disposed to remain on the best terms with those whom they carry. Why, then, the committee would ask, cannot or should not this principle of self-interest, as universal as the existence of man, be adopted in the transportation of passengers generally? As the law stood formerly, wages could not be demanded in case of shipwreck; and, even now, it must be made to appear that every effort has been made by the officers and crew to save the ship, before a claim for wages can be sustained.

Is there any sufficient reason why similar responsibility should not be thrown upon those who have the entire management of the ship, and the means of keeping it in proper condition? Should merchandise be on a better footing than human life? And if there be any propriety or justice in the old common law maxim, that freight was the mother of wages, why should not the safe delivery of the passenger at the port of destination be the foundation of any claim for passage money? The committee can see none, and have therefore adopted the principle, and regard it as, perhaps, the most important feature in the bill which they recommend to be enacted into a law. If they are not very much mistaken, the change here made in the relation between the shipper and the passenger will bring about an entirely new era in the history of passenger ships; as it will be the interest of shippers to land their passengers in safety, so it will be their pleasure to provide their ships with whatever may be necessary to make them healthy and comfortable. Nor will there hereafter be the same carelessness in taking enfeebled, broken down, or diseased passengers on board, as has heretofore existed; shippers will be cautious in seeing, at all events, that the individual, of whose life, during the passage, he is to become virtually the insurer, is not laboring under serious sickness at the time of embarkation. The adoption of this principle will further have a most powerful influence in creating a rivalry between ships as to which is the cleanest and healthiest; passengers will always prefer those which have the reputation of being the most lucky, or in other words, the cleanest, best managed, and affording the greatest comfort to those who embark in them. Nor will the ventilation, disinfection, &c., of these vessels hereafter require the interference of Congress; each ship master will be as desirous to have his ship sweet, wholesome, and pleasant in its arrangements as the hotel keeper who desires to attract patronage by the superior neatness, cleanliness, and good order of his establishment. The owners of steamers on the noble rivers of our country, and stage coaches on its great thoroughfares, require no legislation in regard to the spaciousness and airiness of their apartments, or the good driving, and safety of construction of the other, simply because self-interest dictates to them the comfort and security of the travelling public as the best means to secure custom. So will it be with the owners of passenger ships, when left to themselves; and they will provide for the health and comfort of the poor emigrant as well as of the millionaire, because, in doing so, they will protect their purses, and protect their own interests.

In concluding their report, the committee think proper to present a brief outline of what they would recommend, after a careful examination of the subjects submitted to their consideration.

They deem it proper, and therefore recommend, that a space be reserved on the upper deck, and kept clear, for the enjoyment of air and exercise by the passengers. If this space be not sufficient for the accommodation of all the passengers at once, they may still be divided into squads, and take their exercise in turn at certain hours of each day, should the weather permit. Nothing is more conducive to the cure of sea-sickness and the preservation of the strength and good spirits of passengers at sea than the necessity of moving about and becoming interested in what is going on around them. At the same time, the temporary absence of passengers from the sleeping apartments will afford an opportunity for cleansing them, and removing whatever may be offensive or injurious to health.

They also recommend a restoration of the limitations which existed until the year 1848, of two passengers to every five tons register, but without diminishing the space allowed under the present law to each passenger, in order to prevent the crowding that now takes place of vast numbers in any one ship. It is impossible to ventilate properly an apartment of six or even eight feet in height into which five, six, and sometimes eight or nine hundred persons are crammed; and it is evident, from the information furnished to the committee, that the greatest mortality has prevailed where the ships have been most crowded. One ship from Liverpool for New York, with nine hundred and twenty passengers, lost one hundred in a passage of only thirty-three days.

The committee are of opinion that during the winter the number of passengers should be limited to one for every three tons, as the inclemency of the weather and consequent necessity of keeping the hatches closed obstruct ventilation, and prevent the passengers from taking their exercise in the open air.

The prohibition of passengers on the orlop deck, for the reasons assigned in the preceding part of this report, is earnestly recommended.

An increase of the number of privies, and separate ones for the females, are recommended; the present allowance of one to every hundred passengers is totally inadequate. Common decency would seem to indicate the propriety, or rather necessity, of having separate accommodations for females, whose health and comfort are often destroyed by the fear of exposure to the brutal remarks of the vulgar and obscene among the male passengers and the crew.

As the committee believe that it is entirely impossible that hundreds of people, some of them feeble from sickness, can do their cooking at a caboose four feet by one and a half in size, they have recommended that the provisions shall be cooked as well as furnished by the ship. The reasons for this arrangement are given more in detail in a preceding part of this report. The Bremen ships, so celebrated for the general good condition of their passengers, adopt this course, and find it to work admirably.

With a view to confer upon the captain ample power to maintain discipline without being subjected, as at present, to vexatious prosecutions, the committee have recommended the adoption and publication of certain rules throughout the ship, the observance of which is to be enforced by the master, in case of resistance to his proper orders and authority; a statement of the facts to be entered at the time on the log, and signed by the surgeon (if any be on board) and mate, and read to the offender; this statement, so made, to be available as *prima facie* evidence of justification in all suits or complaints brought or made against him, or against any who aid in carrying out his orders.

Believing that by making the shipper virtually the insurer of the passenger's life during the passage, will have a most salutary tendency, the committee, for the reasons heretofore assigned at length, recommend that the ship be made responsible to the extent of the passage money in the event of death during the passage.

A more accurate return of the names and descriptions of passengers and of the deaths on board is required, and is provided for by the bill which is reported herewith.

The committee have deemed it proper to annex, in the form of an appendix, some of the communications addressed to them. They would have been most happy to have attached to their report the whole mass of documents laid before them, as all are valuable, but to have done so would have been to extend the report to an extreme bulk. The communications selected, contain all or very nearly all of the views expressed in those which have been omitted, the principal difference being in the form in which they are set forth.

The committee are under obligations to the Hon. W. L. Marcy, Secretary of State of the United States, for a copy of a very able and lucid communication from the Hon. A. Dudley Mann, on the subject of emigration, prepared by the latter, under instructions from Mr. Buchanan, while Secretary of State, which forms a part of the appendix. This document is entitled to and will be received with respect; it needs no commendation at the hands of the committee. The following papers are communicated herewith:

No. 1. Circular of the committee.

No. 2. Communication from the Hon. A. Dudley Mann, to Hon. James Buchanan.

No. 3. Communication from Samuel Hall, of East Boston.

No. 4. Communication from John H. Griscom, M. D., of New York, to the Special Committee of the United States Senate.

No. 5. Communication from Gregory Dillon, esquire, President of the Irish Emigration Society of New York.

No. 6. Communication from the German Society of New York.

No. 7. Communication from Leopold Bierwirth, esquire in answer to the circular of the committee.

No. 8. Letter from R. B. Minturn, in answer to circular of the committee, and enclosing letters from Captains Knight and Britton, experienced commanders of passenger ships.

No. 9. Letter from Cyrus Curtis, esquire, one of the commissioners of emigration of New York.

No. 10. Letter from Messrs. Oelricks and Company.

No. 11. Paper received from Mr. Rucker, late minister from Hamburg to Prussia.

No. 12. Letter from Messrs. Meyer and Slucken, owners of passenger ships.

No. 13. Letter from Adolf Rodewald, esquire.

No. 14. Letter from Captain Wm. Skiddy.

No. 15. Letter from E. D. Hurlbut, esquire.

No. 16. Letter from Dr. Isaac Wood, of New York.

No. 17. Communication from the Board of Health, of Philadelphia.

No. 18. Letters from A. Schumacker, esquire, President of the German Society of Baltimore, in one of which a draught of a law to regulate passenger ships, is enclosed.

No. 19. Report of the Board of Trade of Baltimore.

No. 20. Communication of Dr. Cartwright, of New Orleans.

No. 21. Communication from the Mayor of New Orleans.

The committee respectfully recommend the passage of the accompanying bill.

A.

Table showing the number of passengers arrived in the United States from foreign ports from October 1, 1843, *to December* 31, 1853, *as reported annually to Congress by the Secretary of State, under the act of* 1819.

Year ending	Males.	Females.	Sex not stated.	Total.
September 30, 1844	48,897	35,867	....	84,764
Do. 1845	69,188	49,290	1,406	119,884
Do. 1846	90,973	66,778	897	158,648
Do. 1847	139,166	99,325	989	239,480
Do. 1848	136,128	92,883	472	229,483
Do. 1849	179,253	119,915	442	299,610
Do. 1850	200,903	113,392	1,038	315,333
September 30 to December 31, 1850	38,282	27,107	181	65,570
Year 1851	245,017	163,745	66	408,828
1852*	....	....	....	398,470
1853	236,596	164,181	....	400,777
				2,720,847

*The report of the Secretary of State to Congress for this year does not give the number of each sex.

B.

Table showing the number of passengers on vessels arriving at the ports of Boston, New York, Philadelphia, Baltimore, Charleston, and New Orleans during the year 1853, *and the number of deaths: compiled from the returns made to the committee by the collectors at those ports.*

1853.	Boston.		New York.			Philadelphia.		Baltimore.		Charleston.*		New Orleans.	
	Passengers.	Deaths.	Passengers.	Deaths.	Landed sick.	Passengers.	Deaths.	Passengers.	Deaths.	Passengers.	Deaths.	Passengers.	Deaths.
January	652	11	4,280	29	50	528	4	212	....	1	....	5,683	3
February	480	1	11,699	129	141	344	....	547	....	31	....	1,289	16
March	560	1	9,381	13	67	1,266	....	415	1	32	....	3,680	24
April	2,664	8	23,045	47	96	1,386	....	134	....	2	....	3,775	2
May	3,143	7	34,596	73	137	2,255	6	307	1	....	....	3,527	12
June	3,657	5	40,009	35	105	3,023	3	1,887	7	28	....	4,214	23
July	1,602	2	21,276	46	55	1,462	1	532	....	....	....	789	3
August	4,060	2	34,270	....	66	1,774	....	2,172	4	35	....	64	1
September	3,224	18	28,652	280	107	2,017	7	1,457	2	....	....	219	....
October	1,875	70	21,499	277	80	2,111	4	1,507	3	....	....	769	6
November	3,399	2	31,191	1,112	241	1,476	....	655	....	420	1	5,194	36
December	731	....	16,551	177	75	971	7	1,444	37	50	....	8,722	185
	26,047	†127	276,449	‡2,218	1,220	18,613	32	11,269	55	599	1	37,925	311

* Collector's return professes only to give the number of "emigrant passengers."

† The superintendent of alien passengers gives 146 as the amount of deaths on board of "emigrant ships" alone arriving at Boston. Of these, 80 occurred on board of the Sagadohock and Lexington, the first of which lost 62 and the latter 18, from Gottenburg; and were caused, as he says, "by improper food with which the passengers supplied themselves," and also "the use of water put up in old oil casks."

‡ The number of deaths here given is taken from the report of the collector. According to the statement of the health officer, who is appointed under the authority of the State of New York, the number is much greater, being 3,475. The number of those landed sick is from the report of the health officer, as the collector makes no return of the kind. The collector of the port of New York, in transmitting his statement, says: "I cannot vouch for the correctness of the list in regard to the number of deaths, as I am told by the boarding officer at Staten Island that the reports of the masters are often very contradictory and uncertain. It seems they frequently report to the doctor a much smaller number of deaths than they afterwards admit to the boarding officer; and that even the latter is often less than the number reported by passengers."

The report of the Secretary of State to Congress, under the act of 1819, made in March, 1854, (Ex. Doc. No. 78, 1st sess., 33d Congress,) gives the following as the number of passengers arriving at each of the above named ports during the year ending December 31, 1853: Boston, 25,832; New York, 294,818; Philadelphia, 19,211; Baltimore, 11,368; Charleston, 1,068; New Orleans, 43,028. Total, 395,325.

SUMMARY.

From collectors' returns to committee.

	Passengers.	Deaths.	Landed sick.	From Secretary State's report to Congress.
Boston	26,047	127	....	25,838
New York	276,449	2,218	1,220	294,818
Philadelphia	18,613	32	....	19,211
Baltimore	11,269	55	....	11,368
Charleston	599	1	....	1,068
New Orleans	37,925	311	....	43,028
	370,902	2,444	1,220	395,325

C.

Table showing the number of vessels arriving with passengers at the port of New York from European ports during the last four months of the year 1853, with the number of passengers, deaths, &c.

SEPTEMBER, 1853.

From—	Number of vessels.	Number of passengers.	Average of passengers.	Average length of passage.	Number of deaths by collector's report.	Number of deaths by health officer's report.*	Number landed sick.	Per centage of deaths.	Per centage of deaths, and landed sick.	Number of vessels with cholera.	Number of passengers in those vessels.	Average of passengers.	Average length of passage.	Number of deaths.	Number landed sick.	Per centage of deaths.	Per centage of deaths, and landed sick.	Number of ships on which deaths occurred, not reported as cholera.	Number of passengers in those vessels.	Average number of passengers.	Average length of passage.	Number of deaths on those vessels.	Number landed sick from those vessels.
				Days.									*Days.*								*Days.*		
Liverpool	39	15,065	386	39	225	207	66	1.49	1.93	4	2,309	577	33	225	33	9.74	11.17	......	......	......	......	......	......
London	6	1,567	261	36	......	9	3	......	.19	......	......	......	......	......	......	......	......	......	......	......	......	......	......
Bremen	16	3,743	232	44	49	64	15	1.31	1.72	1	330	330	41	48	12	14.54	18.18	1	258	258	40	1	......
Hamburg	6	1,398	233	39	16	23	10	1.14	1.86	1	287	287	40	16	6	5.57	7.66	......	......	......	......	......	......
Havre	8	3,441	430	34	......	11	3	......	.08	......	......	......	......	......	......	......	......	......	......	......	......	......	......
Other European ports	19	2,458	129	41	......	8	......	......	......	......	......	......	......	......	......	......	......	......	......	......	......	......	......
	94	27,672	294	39	290	322	97	1.04	1.89	6	2,926	488	35	289	51	9.87	11.62	1	258	258	40	1	......

* This column gives the number of deaths reported by the "health officer," who is the physician appointed by the State authorities to board and inspect vessels on their arrival at quarantine. Although this report is very possibly more accurate than the returns made by the captains to the collector of the port, (given in the preceding column,) the report of the collector, he being an officer of the United States, and the returns to him being made under an act of Congress, is taken as the basis of all the statements in these tables.

C—Continued.

OCTOBER, 1853.

From—	Number of vessels.	Number of passengers.	Average of passengers.	Average length of passage.	Number of deaths by collector's report.	Number of deaths by health officer's report.	Number landed sick.	Per centage of deaths.	Per centage of deaths and landed sick.	Number of vessels with cholera.	Number of passengers in those vessels.	Average of passengers.	Average length of passage.	Number of deaths.	Number landed sick.	Per centage of deaths.	Per centage of deaths and landed sick.	Number of ships on which deaths occurred, not reported as cholera.	Number of passengers in those vessels.	Average number of passengers.	Average length of passage.	Number of deaths on those vessels.	Number landed sick from those vessels.
				Days.									*Days.*								*Days.*		
Liverpool	27	11,719	434	44	233	256	63	1.98	2.53	6	3,269	545	40	213	32	6.51	7.49	5	2,268	453	45	20	1
London	4	1,317	329	43		13	3		.02														
Bremen	10	2,055	205	49	4	16	2	.02	.03									2	616	308	47	4	1
Hamburg	5	975	195	44	20	36		2.05	2.05	*3	548	183	44	20		3.65	3.65						
Havre	8	3,056	381	41	1	16	5		.02									1	356	356	48	1	
Other European ports	16	2,113	132	52	2	9	1	.01	.01									1	232	232	49	2	
	70	21,229	307	46	260	346	74	1.22	1.57	9	3,817	424	41	233	32	6.10	6.96	9	3,472	386	47	27	2

* The Hamburg ship Copernicus is not included in this. She arrived with 207 passengers, after 55 days passage, no death being reported by the collector; but the "health officer" reports 15 deaths from cholera.

C—Continued.

NOVEMBER, 1853.

From—	Number of vessels.	Number of passengers.	Average of passengers.	Average length of passage.	Number of deaths by collector's report.	Number of deaths by health officer's report.	Number landed sick.	Per centage of deaths.	Per centage of deaths and landed sick.	Number of vessels with cholera.	Number of passengers in those vessels.	Average of passengers.	Average length of passage.	Number of deaths.	Number landed sick.	Per centage of deaths.	Per centage of deaths and landed sick.	No. of ships on which deaths occurred not reported as cholera.	Number of passengers on those vessels.	Average number of passengers.	Average length of passage.	Number of deaths on those vessels.	Number landed sick from those vessels.
				Days.									*Days.*								*Days.*		
Liverpool	28	14,760	527	39	753	730	130	5.10	5.97	14	8,264	590	41	731	101	8.84	10.09	5	2,309	461	39	22	24
London	4	1,624	406	41	100	100	16	6.15	7.16	3	1,282	427	38	100	16	7.80	9.05	.....	.....	.....	.....	.....	.....
Bremen	15	2,732	182	46	20	41	8	0.73	1.02	1	236	236	37	15	6	6.36	8.89	3	722	240	43	5	.....
Hamburg	7	1,252	179	45	46	54	11	3.67	4.55	2	426	213	39	40	7	9.39	11.03	3	280	140	48	6	.....
Havre	15	6,678	447	39	217	239	26	3.25	3.64	4	2,132	533	40	213	25	10.00	11.11	2	1,130	565	28	4	.....
Other European ports	22	3,958	180	43	70	59	29	1.77	2.50	3	555	185	48	56	11	10.01	12.07	3	830	271	38	14	.....
	91	31,004	340	42	1,206	1,223	220	3.88	4.60	27	12,895	478	41	1,155	166	8.97	10.26	16	5,271	330	40	51	24

DECEMBER, 1853.

From—	Number of vessels.	Number of passengers.	Average of passengers.	Average length of passage.	Number of deaths by collector's report.	Number of deaths by health officer's report.	Number landed sick.	Per centage of deaths.	Per centage of deaths and landed sick.	Number of vessels with cholera.	Number of passengers in those vessels.	Average of passengers.	Average length of passage.	Number of deaths.	Number landed sick.	Per centage of deaths.	Per centage of deaths and landed sick.	No. of ships on which deaths occurred not reported as cholera.	Number of passengers on those vessels.	Average number of passengers.	Average length of passage.	Number of deaths on those vessels.	Number landed sick from those vessels.
Liverpool	18	7,156	397	33	7	13	9	.10	.22	.....	.....	.....	.....	.....	.....	.....	.....	3	1,415	471	30	7	1
London	6	2,016	336	34	78	70	9	3.86	4.36	2	774	387	37	63	9	8.14	9.30	1	42	42	42	15	9
Bremen	11	1,930	175	45	27	76	12	1.40	2.03	1*	212	212	40	25	.....	13.35	13.35	2	376	188	48	2	3
Hamburg	4	961	240	38	9	17	7	.93	1.65	.....	.....	.....	.....	.....	.....	.....	.....	1	210	210	45	9	5
Havre	11	4,013	365	35	56	82	29	1.39	2.12	2	1,233	616	30	56	26	4.54	6.65	.....	.....	.....	.....	.....	.....
Other European ports	7	969	188	42	.....	9	6	.....	.....	.....	.....	.....	.....	.....	.....	.....	.....	.....	.....	.....	.....	.....	.....
	57	17,045	299	37	177	267	66	1.56	1.99	5	2,219	445	35	144	35	6.49	8.07	7	2,043	292	39	33	18

* The "Nelson" is not included in the above. She arrived December 15, 1853, after forty-nine days' passage, with 351 passengers. The collector reports no deaths; but the health officer reports 26 deaths by cholera, and 9 landed sick; also, on the 14th of December, the "Union" arrived, after forty-three days' passage, with 232 passengers. The collector reports no deaths; but the health officer reports 15 deaths by cholera.

C—Continued.

Summary of the preceding tables, showing the arrivals, &c., at the port of New York from different Europeans ports, from September 1 to December 31, 1853, inclusive.

From—	Number of vessels.	Number of passengers.	Average of passengers.	Average length of passage.	Number of deaths by the collector's report.	Number of deaths by the health officer's report.	Number landed sick.	Per centage of deaths.	Per centage of deaths and landed sick.	Number of vessels with cholera.	Number of passengers in those vessels.	Average of passengers.	Average length of passage.	Number of deaths.	Number landed sick.	Per centage of deaths.	Per centage of deaths and landed sick.	No. of vessels in which deaths occurred not reported as cholera.	Number of passengers in these vessels.	Average number of passengers.	Average length of passage.	Number of deaths in these vessels.	Number landed sick from these vessels.
				Days.									*Days.*								*Days.*		
Liverpool	112	48,700	435	39	1,218	1,206	268	2.50	3.05	24	13,842	577	38	1,169	166	8.44	9.64	13	5,992	461	39	49	26
London	20	6,524	326	38	178	192	31	2.55	3.05	5	2,056	411	38	163	25	7.92	9.14	1	42	42	42	15	9
Bremen............	52	10,460	201	46	100	197	37	.95	1.31	3	778	259	39	88	18	11.29	13.62	8	1,972	246	45	12	4
Hamburgh	22	4,586	208	42	91	130	28	1.94	2.59	6	1,261	210	41	76	13	6.03	6.90	4	490	122	47	15	5
Havre.	42	17,182	409	38	274	348	63	1.59	1.95	6	3,365	561	36	269	51	7.98	9.51	3	1,486	495	38	5	.....
Other Europ'n ports.	64	9,498	148	41	72	85	30	.76	1.07	3	555	185	48	56	11	10.09	12.07	4	1,062	265	41	16	.....
	312	96,950	311	41	1,933	2,158	457	1.99	2.46	47	21,857	465	39	1,821	284	8.42	9.68	33	11,044	335	40	112	44

Summary of the preceding table, showing the aggregate arrivals, &c., in the port of New York from European ports, during each of the last four months of 1853.

	Number of vessels.	Number of passengers.	Average of passengers.	Average length of passage.	Number of deaths by the collector's report.	Number of deaths by the health officer's report.	Number landed sick.	Per centage of deaths.	Per centage of deaths and landed sick.	Number of vessels with cholera.	Number of passengers in those vessels.	Average of passengers.	Average length of passage.	Number of deaths.	Number landed sick.	Per centage of deaths.	Per centage of deaths and landed sick.	No. of vessels in which deaths occurred not reported as cholera.	Number of passengers in these vessels.	Average number of passengers.	Average length of passage.	Number of deaths in these vessels.	Number landed sick from these vessels.
September	94	27,672	294	40	290	322	97	1.04	1.89	6	2,926	488	35	289	51	9.87	11.62	1	258	258	40	1	.....
October	70	21,229	307	48	260	346	74	1.22	1.57	9	3,817	424	41	233	32	6.10	6.96	9	3,472	386	47	27	2
November	91	31,004	340	42	1,206	1,223	220	3.88	4.60	27	12,895	478	41	1,155	166	8.97	10.26	16	5,271	330	41	51	24
December.	57	17,045	299	38	177	267	66	1.56	1.99	5	2,219	445	35	144	35	6.49	8.07	7	2,043	292	39	33	18
	312	96,950	311	41	1,933	2,158	457	1.99	2.46	47	21,857	465	39	1,821	284	8.42	9.68	33	11,044	335	40	112	44

No. 1.

SENATE CHAMBER,
Washington, December 29, 1853.

SIR: A Select Committee, appointed by the Senate of the United States, to inquire into the causes and the extent of the sickness and mortality prevailing on board the emigrant ships, have instructed me to obtain the opinion of gentlemen of experience and of professional knowledge, both with reference to any deficiency in the provisions of the existing statutes, and the propriety of further legislation.

In conformity with their direction, I take leave to ask your opinion as to the adequacy of the existing laws with respect to—

I. The space allotted to each passenger.

II. The quantity and the quality of the provisions required for each passenger.

III. The permission allowed to the passengers to furnish their own provisions for the voyage, instead of making it, in all cases, the duty of the master to provide them.

IV. Ventilation.

V. The cooking arrangements.

VI. The duty of the master to enforce personal cleanliness, and to insure the cleanliness of the vessel.

The committee further request your opinion as to the propriety of amending the existing laws by requiring—

VII. The employment of a qualified and experienced surgeon.

VIII. The employment of a reasonable number of attendants to minister to the sick, and to enforce the observance of cleanliness, both of the persons of the passengers and of the vessel.

IX. The separation of the sexes; and the prevention of unnecessary intercourse between the crew and the passengers.

X. A thorough process of disinfecting every vessel, on board of which disease has once made its appearance.

XI. A report to be made by every vessel bringing emigrant passengers, of the length of voyage, number of passengers, number of deaths, &c., to be published, and to be returned to the State Department.

XII. In case deaths have occurred during the voyage, an inquest to be held under the supervision of federal officers, and the verdict to be published and returned as above.

XIII. A limitation to the number of passengers allowed in any vessel, in proportion to the tonnage of the vessel.

XIV. A distinction with respect to the number of passengers between vessels passing within the tropics, and those not so passing.

The committee will also be happy to receive from you any statement of facts within your knowledge tending to exhibit the extent, or the causes of the sickness and mortality which have prevailed, or the insufficiency of the provisions of the existing laws, as well as any suggestions which you may think proper to make in connection therewith, or with regard to the proper remedy to be applied.

An early reply, with answers to any or all of the points above suggested, will be esteemed a favor.

Very respectfully,
HAMILTON FISH,
Chairman.

NOTE.—The principal provisions of the existing laws on the subject will be found in chapter 16, Laws, 2d session 29th Congress, (Feb. 22, 1847;) chapter 41, Laws, 1st session 30th Congress, (May 17, 1848;) chapter 111, Laws, 2d session 30th Congress, (March 3, 1849,) and chapter 46, Laws, 2d session 15th Congress, (March 2, 1819.)

No. 2.

Communication from Hon. A. Dudley Mann, to Hon. James Buchanan.

BREMEN, GERMANY,
September 13, 1847.

SIR: Having visited the ports of western Europe, designated in your instructions dated June 26, I now proceed to communicate to you all the information, of a valuable nature, which I have collected, relating to the subject of emigration.

In conformity with the act of parliament, entitled "an act for regulating the carriage of passengers in merchant vessels" approved August 12, 1842, there appears to be entire uniformity at the ports of the United Kingdom of Great Britain and Ireland, in the rules and regulations observed for the embarcation, and conveyance of emigrant passengers. I deem it important to make a somewhat lengthened abstract from this enactment, as far as it has a bearing upon voyages to North America.

Under its provisions, vessels are permitted to carry three persons (officers and crew included) to every five tons burthen, provided that there be not more than one person to every ten clear superficial feet of the space intended for the use of passengers. Two children under the age of fourteen years are computed as one passenger, and children not above one year are not taken into the estimate. For every passenger carried by any vessel beyond this proportion, a penalty of five pounds is exacted.

The lower deck of vessels is required to be of the thickness of an inch and a half, and made secure to the hold beams. The height between decks must be six feet: There must not be a greater number than two tiers of berths—the bottom of the lower tier to be six inches above the deck: The berths to be six feet long and eighteen inches wide, and one allotted to each passenger.

A supply of food, at the rate of one pound of breadstuffs per diem—half at least to be bread or biscuit, the remainder may be potatoes, five pounds of which must be estimated as equal to one pound of bread, must be issued, not less frequent than twice a week to each passenger, as likewise three quarts of water to each, daily. Breadstuffs and water must be inspected and surveyed by the government emigration agents, or, in their absence, by the officers of customs, who are also to enquire into and determine upon the sea-worthiness of vessels ordering a survey to ascertain their true condition, if necessary. The length of a voyage must invariable be computed at seventy days; and a sufficiency of boats are to be taken for any casualty that may occur.

An ample supply of medicines to be provided for the voyage, and suitable directions prepared for their use; no ardent spirits are permitted to be sold, under a penalty of one hundred pounds. During the detention of a vessel, her passengers must be victualled; and where a delay exceeds two clear working days, unless where it is occasioned by wind or weather, the passengers are to receive instead one shilling per diem, or the contractor may, with their consent, lodge and maintain them. Passengers shall not be landed, except they request it, at any place other than the port which they sailed for; nor are they compelled to disembark for forty-eight hours after reaching their destination, unless the ship finds it necessary, in continuation of her voyage, to leave the port before that time elapses.

No person, except the owners or masters of vessels, can act as a passage broker or dealer, unless duly licensed by magistrates, under a penalty of ten pounds for every offence; and the same penalty is imposed upon brokers, with a forfeiture of their license, if they resort to fraudulent means of inducing persons to engage passage.

Where the contract for a passage is not complied with by the shipper, the party aggrieved may recover, by summary process, before two justices of the peace, any passage money which he has paid, with a sum not exceeding fifty pounds as a remuneration, provided he is not maintained at the expense of the shipper, and a passage secured for him within a reasonable time. Parties covenanting for passages must give written receipts, in a prescribed form, for money received, under a penalty not exceeding ten pounds, and if a passage broker, a forfeiture of license.

Passengers instituting suit against shippers, for money made recoverable for their benefit, shall not be rendered, on that account, incompetent witnesses in the case. Their right, by an action at law, for a breach of contract, is reserved.

Masters of vessels must afford all necessary facilities to the authorized officers for inspecting the ship, and communicating with the passengers, with a view to the proper discharge of their official duties. A penalty, not to exceed fifty pounds, may be imposed for the breach of any of the provisions of the law, other than in the cases where specific penalties are affixed.

Foreign as well as British ships are included in the provisions of the act; but it is almost needless for me to remark, that in their intercourse with the ports of the United States, they all have to be governed by our legislation, with respect to the number of passengers which they are entitled to carry. The acts of Congress, approved February 22 and March 2 of the present year, upon the subject, were promptly and properly explained, and made sufficiently public, in Great Britain, by "her majesty's colonial land and emigration commissioners."

By an act of Parliament, "to amend the passengers act, and to make further provision for the carriage of passengers by sea," approved July 22, 1847, it is provided that the colonial land and emigration commissioners may, in certain cases, substitute other food in lieu of the description required by the law of 1842, but they shall not increase the per diem allowance; the food so substituted shall be laden on board the vessel, at the expense of the owner or charterer, and shall be of a quality to be approved by the emigration officer at the port of clearance; and that in default, the owner, charterer, or master, shall be liable to a penalty of fifty pounds.

No ship carrying passengers shall be allowed to take, as cargo, any gunpowder, vitriol, or green hides.

Passengers shall, at all times during the voyage, (weather permitting,) have free access to and from the between decks, by each hatchway situate over the space appropriated to their use; and if the main-hatchway be not one of the hatchways appropriated for such purpose, or if the natural supply of light and air through it be unduly impeded, the emigration officer may direct such improvements to be made as will ensure a sufficiency of light and air for the comfort of the passengers.

Ships proposing to carry passengers shall, when a demand is made by the emigration officer at the port of clearance, be surveyed by two or more competent surveyors, duly authorized and approved of, either by the commissioners of colonial lands and emigration or by the commissioners of customs; and if such surveyors shall report adversely to her sea-worthiness, she shall not be cleared out until she has been rendered in all respects fit for her intended voyage. This provision "shall in all respects, and without distinction, be the same for foreign as British ships." And furthermore, a clearance shall not be granted to such ship until it has been satisfactorily ascertained by the proper authority that she is manned with a full complement of men; nor shall she be allowed to proceed on a voyage until she obtains a certificate from the emigration officer at the port of clearance—or his assistant, or, in the absence of both, from the officer of the customs—that all the requirements of the present law, as well as of the law of 1842, have been duly complied with.

If a ship having passengers on board shall put into any port of the United Kingdom after sailing, she shall replenish her stock of food, water, and medical stores, so as to make up the full quantities of those articles, of the requisite kind and qualities. In default thereof, the master shall incur a penalty not exceeding one hundred pounds.

In case any ship carrying passengers shall be wrecked or otherwise destroyed, or shall put into a port or place in a damaged state, or be prevented otherwise from landing her passengers at their port of destination, she shall provide them with a passage in some other equally eligible vessel; and if the master should fail, in a reasonable time, to comply with this requisition, they shall be entitled to recover, by summary process, before any two justices of the peace, from the owners, charterers, or master of such ship all moneys which shall have been paid by or on account of them; together with a further sum, not exceeding five pounds, for each passenger, as shall, in the opinion of the two justices, compensate for the loss or inconvenience occasioned by the delay.

In order to remove the ambiguity of the act of 1842, respecting the number of passengers which a vessel may rightfully carry from a port of Great Britain, it is provided that, in reckoning the proportion of passengers to tonnage, two children under the age of fourteen years shall be computed as one passenger, and children under one year old shall not be counted.

Vessels may carry one passenger to every twenty-five tons of their measurement, without being subjected to the provisions of the acts of Parliament, relative to passenger ships, except for the refunding, in certain cases, passage money, &c.

The aforegoing epitome will, perhaps, make you sufficiently acquainted with the laws of Great Britain concerning emigration, but I nevertheless consider it necessary, for the purpose of reference, to furnish you with copies of the original enactments, and accordingly I herewith transmit them. The enforcement of the provisions of these acts devolves, exclusively, upon the "emigration officers at London, Liverpool, Plymouth, Glasgow, Dublin, Cork, Belfast, Limerick, Sligo, Donnegal, Londonderry, Waterford, Baltimore, Galway, Newry, and Tralee, and at all other ports upon the 'officers of the customs.'"

It is their duty to procure and give information, when desired, gratuitously, as to the time of the sailing of ships, the price of passage, the outfits and stores of provision indispensable for a voyage, and the regultaions to be observed by passengers during the period they are at sea. The emigration officers are selected from the lieutenants of the royal navy, and act under the immediate directions of the colonial land and emigration commissioners, by whom they are instructed to urge upon captains and passengers a faithful observance of the following rules:

All the emigrants to be out of bed by seven o'clock in the morning, and the children to be regularly washed and dressed. The fires to be lighted by seven a. m., and to be extinguished by seven p. m., unless the captain should direct otherwise. When the emigrants are dressed, the beds to be rolled up, and, weather permitting, taken on deck. They are to be well shaken and aired at least twice a week. The bottom boards of the berths should, if not fixtures, be removed and dry scrubbed, and taken on deck once or twice a week. The decks to be swept before breakfast, and after every meal; including in the first sweeping the space under the bottom of the berths, the boards of which, if not fixtures, are to be lifted for the purpose. The decks to be also dry holystoned or scraped at nine a. m., and the occupants of each berth are to see that it be well brushed out, and the space in front made clean. All emigrants to be in bed by ten o'clock.

A party of six, or more, to be formed from all the adult males, above fourteen, taken in rotation, with the exception of the overseers, to clean such parts of the deck as do not belong to any particular berths, and also the ladders and water-closets, and hospitals, if there be any, and to be sweepers for the day. The coppers and cooking utensils to be cleaned daily.

When the ship undertakes wholly to victual the emigrants, they should be divided into messes of not less than six nor more than twelve adults, and one or two men should be taken in turn to cook daily for all the messes. When the emigrants victual and cook for themselves, the overseers should see that each family has its regular turn at the fire place.

The issues of bread-stuffs, required by the "passengers act" of 1842, should be made by an officer of the ship in the presence of the overseers. They are intended merely to prevent destitution during the voyage; and further supplies, consequently, should be put on board by the passengers themselves if they do not make an arrangement with the ship to victual them beyond the legal allowance. Water to be issued also by an officer of the ship. Medicine to be had, when wanted, by applying to the captain.

Washing days Monday and Friday, or such other days as the captain may find it more convenient to appoint, with reference to weather and other circumstances, but no washing or drying of wet clothes to be suffered between decks on any pretence whatever. On every Sunday, at half past ten, the passengers will be assembled, and it is expected that each of them will then appear clean, and put on clean linen and decent apparel. The Lord's day is to be as religiously observed as circumstances will admit. Swearing and all improper language strictly forbidden. No fighting, riotous or quarrelsome behavior allowed.

In cases of disputes, the matter should be at once submitted for the decison of the captain, whose authority is to be duly respected, and swords, pistols, and other arms must be placed in his charge. No gambling permitted.

No passengers allowed to go to the ship's cook-house without a permit from the captain or chief officer, or to be in the forecastle, among the sailors, on any account; nor are sailors suffered to be in the between decks amongst the passengers, except on duty.

No spirits or gunpowder allowed to be brought on board. If discovered, they will be taken from the party having them. No smoking allowed between decks, and no loose hay or straw permitted to be taken below.

A lamp to be kept burning all night at the main hatchway, but no other light after eight o'clock, p. m. No naked light allowed below on any account.

Since the disease usually called "ship fever" manifested itself, the emigration officers require passengers, before they proceed to sea, to submit to a medical examination, and such as are ascertained to be ill are prevented from departing. The fever originated from a want of nourishment in Ireland, last autumn—which numbered its hundreds of thousands of victims—is the identical malady that has produced such distressing mortality among the emigrants embarking from the ports of Great Britain for our shores, and for the British possessions on our continent. This is the unanimous belief of the emigration officers, and the general one of physicians and others in Ireland. Its correctness is corroborated by indisputable facts, one of which it may not be amiss to relate. On the 15th of March last, a vessel sailed from Liverpool, with 296 Irish emigrants from the county of Roscommon, for Philadelphia. On the 17th of the same month, having in the mean time encountered a severe gale, in which she lost her foremast, she put into

Belfast, where she landed 56 passengers, who were prostrate with "ship fever," many of whom subsequently died. On the 17th of April, after undergoing repairs, she again sailed, but soon losing her main and mizen masts, entered Londonderry, in a distressed condition. There 24 of her passengers died, and ten were left in the hospital, dangerously sick. The captain, the crew, and a surgeon, who was aboard, contracted the disease, on which account the voyage could not be resumed until the 20th of July. Starvelings conveyed the distemper, engendered by destitution, to this vessel, and although the fever is of a less contagious character than some others, it communicated itself, in a space so circumscribed as a "between decks," with frightful rapidity to persons utterly free from it.

To one travelling like myself, through the populous cities and districts of Ireland, even after the demon of starvation had been checked in his desolating career, by an abundant importation of supplies from our ever productive fields, there could be nothing wonderful, in view of the coming winter and the probable return of the melancholy occurrences of the past, that all who could should seek a distant asylum. There is nothing unnatural in the desire of the unfortunate Irish to abandon their cheerless and damp cottages, and to crawl inch by inch, while they have yet a little strength, from the graves which apparently yawn for their bodies. What will poor humanity, tenacious of life, not do to avert the blow which death seems to aim? A case was related to me at Dublin of a young wife, many of whose family and friends had perished with hunger, which answers the interrogatory in emphatic language. She had gone aboard the bark which was to convey her across the waters, and the night before the morning of sailing she became a mother. The circumstance was communicated to the emigration officer of the port, who immediately proposed to carry her ashore and provide amply for her wants, until she should be restored to her accustomed health and strength; but she promptly declined the benevolent offer, remarking, that "she would rather start with the hope, however delusive it might prove, of arriving soon in America, than to remain under the positive certainty of a speedy recovery." In fleeing from famine and its concomitant pestilence all look for relief beyond the Atlantic, and those who, by patient toil, have accumulated a few pounds, to plenty and contentment. It has repeatedly occurred, notwithstanding the well intended efforts of the examining physicians, that persons embarked with the disease when no evidences of its existence could be found upon them, and in consequence have not only become victims themselves but have rendered their fellow passengers so likewise. But the fever, in most instances, is, doubtless, conveyed on ship board by another agency. In all Catholic countries there is an extraordinary degree of devotion—proceeding from the best impulses of the human heart—to the memory of the departed; and nowhere, perhaps, does this sentiment prevail to a greater extent than in Ireland. It is indulged in by all classes and conditions. The humblest peasant, so destitute as to be unable to enjoy any other visible remembrance of the lost ones dearest to him, fondly clings to the garments, however ragged and worthless, worn by them when living. From such clothing carried by emigrants, as well as that in use by themselves, an effluvium has arisen as fatal in its tendencies, to all who came in contact with it, as that proceeding from the bodies of those suffering by the disease.

It must not be inferred from the preceding remarks, that I desire to convey the impression that all, or any considerable number of the emigrating Irish, are in indigent circumstances. On the contrary, I was assured at every port I visited, by individuals well informed upon the subject, that in the aggregate those embarking this year compared with former ones, particularly to the United States, contrasted favorably both as to respectability and pecuniary substance. They have been chiefly small farmers, disheartened at the repeated crop failures, and agricultural laborers.

It is estimated upon reliable data, that the sum of one million of dollars, at least, is annually remitted by individuals in the United States, in drafts and "passage certificates," to their relations in Ireland, to enable them to reach our Atlantic ports. It was represented to me by each of the emigration officers with whom I conversed, that impositions were constantly practised, occasioning in most instances much distress, by the vendors of passage certificates upon the purchasers and recipients of them. The accompanying copy of a letter, marked A, addressed by Lieut. Forrest, E. O., at Glasgow, to the Colonial Land and Emigration Commissioners, furnished me by Mr. Walcott, their secretary, with the passage certificate subjoined to it, will explain the nature of the grievances to which they are submitted.

Inasmuch as the agreement is made in the United States, the holder of the certificate is without a remedy for a non-compliance of contract; but I learn that it is intended to revoke the license of any broker to whom a certificate is directed and recognized, unless he promptly provides a passage for the party in whose favor it is drawn. If a similar punishment were inflicted upon the brokers (to whom it would apply) in our cities, the evil would be speedily removed.

The following is a tabular statement of the number of emigrants embarking from the ports of the United Kingdom to all countries, between the 31st of December, 1825, and the 30th of June, 1847, being a period of twenty-two years and a-half:

Years.	U. States.	N. American colonies.	Australia and NewZealand.	All other places.	Total.
1825	5,551	8,471	485	114	14,891
1826	7,063	12,818	913	116	20,910
1827	14,526	12,648	715	114	28,003
1828	12,817	12,084	1,056	135	26,092
1829	15,678	13,307	2,016	197	31,198
1830	24,887	30,574	1,242	204	56,907
1831	23,418	58,067	1,561	114	83,160
1832	32,872	66,339	3,739	196	103,146
1833	29,109	28,808	4,098	517	62,532
1834	33,074	40,060	2,800	288	76,222
1835	26,720	15,573	1,860	325	44,478
1836	37,774	34,226	3,124	293	75,417
1837	36,770	29,884	5,054	326	72,034
1838	14,332	4,577	14,021	292	33,222
1839	33,536	12,658	15,786	227	62,207
1840	40,642	32,293	15,850	1,958	90,743
1841	45,017	38,164	32,625	2,786	118,592
1842	63,852	54,123	8,534	1,835	128,344
1843	28,325	23,518	3,478	1,881	57,202
1844	49,600	22,924	2,229	1,873	76,626
1845	58,538	31,803	830	2,330	93,501
1846	82,289	43,439	2,347	1,826	129,901
1847 (first six months)	101,767	72,281	2,185	1,950	178,183
Total	818,157	698,909	126,548	19,897	1,663,511

It will thus be seen that for the first six months of 1847, the number destined for the United States exceeded that of last year entire, nearly twenty per centum. It is quite certain that the increase for the last six months will not be in the same ratio, yet I feel persuaded that we shall, nevertheless, receive more than double the number of persons from Great Britain in 1847 that we received in 1846.

Formerly, about ninety-five per centum of the British emigrants, proceeding to the United States, embarked at Liverpool. The departures for the first six months of the present year show a dimunition of this proportion of nearly a third. This results from the augmentation of the direct navigation intercourse between the ports of the United States and those of Ireland.

From the answers to my numerous inquiries upon the subject, I am of the opinion that of the emigrants who sailed from Great Britian for the United States, between the first of January last and the thirtieth of June, 75,000 were Irish, 10,000 English, 10,000 Scotch, and 6,767 German.

It is a common practice in Ireland for landlords to contribute pecuniary assistance to their more worthless tenants, in order to get rid of them, to enable them to emigrate to Canada; and I was informed at Limerick that several families had been forwarded in like manner from that port to the United States. This custom, as relates to our country, does not prevail I am confident to any considerable extent, as I could not hear, after strict inquiry, of its having been adopted elsewhere. Nor do I believe that vicious persons or notorious paupers are encouraged by any British functionary connected with the subject of emigration, to embark for the United States. On the contrary, I am satisfied from the interviews which I had with the secretary to the Board of the Colonial Land and Emigration Commissioners and the emigration officers, that embarkations of this description would be prevented rather than countenanced. These officers, without a solitary exception, seem not only to discharge their duties faithfully to their own government, but they are actuated also by motives of exalted humanity in protecting intending emigrants to our shores, as far as possible, against fraud and imposition.

I come now to France; and I regret to state that no laws, rules, or regulations respecting merchant passenger ships exist in this kingdom. France is not an emigrating nation, and those who proceed from Havre to the United States are almost exclusively Germans and Swiss. The business of providing passages is conducted by ship-brokers, ship-agents, and merchants, who charter vessels for the purpose, or pay ship-masters a stated amount for the conveyance of each passenger. No obligation is imposed, except such as may be agreed upon by the parties contracting, for the provisioning of emigrants. I found it impossible to get a statement, which could be relied upon as authentic, of the extent of emigration from Havre for any year previous to the commencement of 1843. Since then it has amounted, annually, to the folling number:

1843	8,553
1844	16,660
1845	23,500
1846	32,381
1847, first six months	16,000
Total	97,094

The emigration through Havre for the first six months of 1847, consisted of 15,250 Germans, 600 Swiss, and 150 French.

Belgium, like France, has comparatively no emigration of her own, and the people of no other country than Germany embark at her ports for the United States. By a decree of King Leopold of March 14, 1843, the "maritime commissioners" is charged with the inspection of vessels engaged in transporting emigrants, reporting upon their condition, and deciding whether they are sea-worthy or not. These officers are also empowered to adopt such measures as they may think necessary to ensure the health and general well-being of passengers during the voyage. If they have reason to believe that there is an infectious or other disease aboard a ship, it is their duty to have an examination made by a naval surgeon in such cases as they suspect, and to prevent those afflicted from embarking. Two passengers only are allowed to be carried to every five tons measurement of the ship, whatever may be her destination. Voyages are computed to any port in the United States, (and must be provided for accordingly,) to continue for ninety days. Ships are not permitted to clear until it has been satisfactorily ascertained that they have aboard the quantity (and of good quality) of stores required by the commissioners, viz: 90 lbs. biscuit, 10 lbs. rice, 10 lbs. flour, 22 lbs. of beans and peas, 16 lbs. salt meat,

6 lbs. butter, 2 lbs. salt, six quarts vinegar, and 40 lbs. coal for each passenger. Potatoes may be substituted, in part, for biscuit. Each passenger must be furnished daily with two-thirds of a gallon of good water; and the vessel must be properly supplied with fresh medicines, with directions for the taking of them, in cases where they may be needed.

Previous to the completion of the railway line from the Rhine to the Scheld, the number of emigrants sailing from Antwerp to ports in the United States was very inconsiderable. The amount of embarkations, annually, for the four years and a half terminating on the last day of June, has been as follows:

1843	2,749
1844	2,961
1845	4,549
1846	11,402
1847, first six months	8,900
Total	20,561

Herewith I send you a copy of the Belgium decree to which I have made reference.

In Holland, under an ordinance of the King, dated December 28, 1837, the substance of which I transmit in a separate form, all ship-owners, ship-brokers, and forwarding merchants located in that kingdom, who wish to forward emigrants, must make a declaration of their intention to the "supervisors of emigration," who are appointed by the authorities of the port. Such declaration must be accompanied by a notarial act, reciting the nature of the obligations imposed on the person making application, to wit: that he shall make provision for the proper reception of the emigrants desiring to pass through Holland at the point at which they enter, and take care that they are well fed and lodged until they arrive at the port at which they are to embark; that he shall provide a good vessel for them, which shall be well provisioned, and have good medical attendance; that she shall be in readiness for their immediate reception, with every necessary outfit for the proposed voyage; that he shall make provision for the maintenance of the emigrants at the port from which they are to embark, in the event of unavoidable delay in the sailing of the ship at the time specified; that in case of sea-damage, shipwreck, or other accident, either in river, on the coast, or in the European seas, he shall put the emigrants aboard another sea-worthy vessel, suitably fitted out with supplies, so that they may not be an expense to the kingdom; that he shall, furthermore, execute a general bond, with such sureties as shall be deemed by the proper authorities good and sufficient, to the amount of one hundred and fifty guilders (sixty dollars United States currency) for each person furnished by him with a passage; that he shall be held responsible for all expenses, damages, and loss of interest which may result from his negligence or those in his service, and also for any breach of good conduct of the emigrants in his charge previous to their departure. The supervisors of emigration are required to see that there is a faithful compliance with the engagements entered into by ship-brokers and others in all that relates to the forwarding of emigrants.

They shall also judge of the sea-worthiness of vessels, and ascertain whether the customary supply of food, water, &c., is aboard. A ship may carry from a port of Holland to any country whose laws will permit, four passengers for every five lasts of her measurement, (about ten tons,) and two children, under fifteen years of age, may be counted as one passenger.

Each passenger must be provided with 50 lbs. bread; 100 lbs. potatoes; 25 lbs. beans and peas; 10 lbs. meal; 30 lbs. rice; 20 lbs. smoked or salted meat; and a proportionate quantity of butter, vinegar, and salt, together with a sufficiency of sweet water. If the voyage is to extend as far as New Orleans, thirty additional pounds of substantial food for every passenger must be put on the vessel. The captain has charge of the provisions and water, which he must distribute daily, in equal quantities, among the passengers. Two hundred pounds of baggage may be taken by an emigrant free of charge. Such clothing only as is necessary for use is permitted to remain in the steerage.

Number of emigrants embarking at Rotterdam for the United States, for the four years and a-half terminating on the thirtieth of June, 1847, inclusive:

1843	1,387
1844	2,143
1845	4,549
1846	5,010
1847, first six months	4,321
Total	17,410

Number embarking during same time at Amsterdam:

1843	261
1844	220
1845	582
1847, first six months	1,226
Total	2,564

Of the emigrants who embarked the first six months of this year from Rotterdam and Amsterdam, 4,000 were Germans, and the remainder Hollanders.

By a decree of the government of Hamburgh of March 26, 1845, the business of forwarding emigrants is limited to citizens and inhabitants of the republic; but masters of foreign vessels may engage passengers, if citizens will become sureties that their engagements will be complied with. Contracts may be made by the individuals themselves who undertake to transport emigrants; but if it is preferable to make them through others, sworn ship-brokers must employed.

The brokers are obliged, before the departure of the passengers, to designate the persons for whom they have made contracts.

The broker, or if none has been employed, the contractor must hand over to the police office before the vessel clears, a statement of all the emigrants embarking; their birth-places, respectively, sex, age, occupation, and destination. The contractor must see that the vessel which he employs for the purpose is in good condition, and conveniently arranged to ensure comfort to her passengers. The steerage, which must be appropriated exclusively to their use, is required to be five feet and a-half high, and divided into berths six feet long. Berths for four adults are to be six feet wide, and not more than two tiers shall be permitted one above the other. No merchandise is permitted to be stowed between the berths of passengers. The contractor must provide a sufficient supply of provisions (if the vessel is proceeding to any port in the United States) for a voyage of thirteen weeks. Each passenger must be furnished, weekly, with two pounds beef, one pound bacon, and five pounds bread, besides dried vegetables such as beans, peas, &c., to the extent of three and a-half pounds, per week. He must also provide ninety gallons of water for each passenger, and a box of suitable medicines, to which there shall be free access. Children under the age of one year are not to be taken into the estimate of passengers; three, between the ages of one and seven years, are to be counted as one adult; and two between the ages of seven and twelve, as one. The contractor must produce to the chief of police an insurance policy, in which it must be stipulated that the company or concern insuring engages to pay all costs which may accrue to the passengers in the event of accidents to the vessel, while she is undergoing repairs, (on account of board, &c.,) and if the vessel shall be so disabled as to prevent her from continuing the voyage, to provide them with a passage to their port of destination. Deserters from the army of any of the States of the German confederacy are not permitted to embark, nor are fugitives from justice. Minors residing in the republic can only embark with the consent of their parents or guardians.

Where a vessel is delayed beyond the time of sailing agreed upon, the contractor must pay each passenger a per diem allowance, (about twenty two cents,) unless he feeds and lodges him. Each passenger has a right to demand a certified copy of the contract made by him, in the German language; and copies of the decree, one of which I herewith transmit, must be pasted up in conspicuous places in every emigrant ship.

Number of emigrants embarking annually at Hamburg for the United States, from the 1st of January, 1836, to the 30th of June, 1847, inclusively:

1836	2,870
1837	2,177
1838	484
1839	1,415
1840	1,720
1841	1,134
1842	495
1483	1,756
1844	1,774
1845	2,388

1846	3,971	
1847 direct	5,317	6,575
6 months, via Quebec	1,258	
Total	26,759	

The Senate of Bremen has, from time to time, passed numerous laws regulating the embarkation and the carriage of emigrants from the Weser. On the 12th of May last the different enactments were embodied into one, and such imperfections as they were believed to contain remedied. The "government decree" I herewith transmit, making as succinct an abstract from its provisions, which relate to the United States, as is necessary to a proper understanding of them.

The right of engaging passengers for a vessel intending to sail for a trans-Atlantic port can be enjoyed only by ship owners, their correspondents, or those at whose disposal a ship is placed by contract for carrying freight and passengers, or passengers alone. Those who forward passengers are required not only to be citizens of the city of Bremen, but also to reside therein and transact business.

Ships are not suffered to be advertised in the public papers of Bremen for passengers, except by their owners, correspondents, the ship-brokers employed by them, and those by whom they have been taken up or chartered.

Ship-brokers of the city of Bremen may themselves engage passengers, or they may have them engaged by agents, under contract to forward at a future period, but upon the express condition that they shall be turned over, with their consent, however, for conveyance across the seas, to a person properly authorized to take charge of them, as before provided for in this enactment. Any one who engages a passenger, or has one engaged by his agent, be it in the territory of the republic or elsewhere, must immediately furnish him with a certificate of engagement, in which his full name, together with the amount of passage money agreed upon, must be stated. Emigrants upon their arrival at the city of Bremen must announce themselves in person at the police office, those going direct to Vegesack or Bremen-Haven to the authorities at those places. The police officers are required, as far as in their power lies, to prevent deserters from any army of the German confederation from embarking.

The 14th section of the decree manifest a spirit of such good faith towards our country that I consider it proper to insert here a literal translation of it. It reads thus:

"Whereas, in the United States of America fears are entertained that the liberty extended to emigrants to take up their abode there is subjected to great abuses by persons fleeing from justice on account of crimes and misdemeanors committed, and those sent thither from European prisons and poor-houses; and, whereas, the friendly intercourse and extended commercial relations which so happily subsist between this republic and the United States, give to the Senate a particular cause for regarding the interests of those States, it is determined that such persons shall not be received on any vessel that may be expedited from Bremen or its ports; and the police authorities are especially enjoined not to permit any such to embark.

"When discovered they shall be arrested and sent back to the State they came from."

Ship-owners and others engaged in the pursuit of forwarding emigrants, are prohibited from receiving, as passengers, persons intended to be embraced in the meaning of the preceeding provision. Should any such be engaged by ship-brokers, or their agents, ignorantly, as soon as they are apprized of the fact they must give information to the director of police. Any one who expedites a ship for a trans-Atlantic port must hand to the "inspection of brokers" a complete list of all the passengers to be dispatched, mentioning the places of their birth, respectively, as well as age, sex, occupation, and destination. This must be accompanied by a declaration that, according to the best of his knowledge, there is no person, whose name is thereon, who is endeavoring to evade the punishment awaiting him (or her) for any crime committed, or who has been sent away from any European house of correction or poor-house, and that he will not allow persons of this description to be received aboard the vessel. Where emigrants have been engaged by a ship-broker, or his agent, the declaration must be accompanied by an oath. If, after a list has been made up, emigrants, whose names are not upon it, apply to the captain for a passage, whether at the embarkation or elsewhere, producing the proper certificates that they are entitled to it, he is authorized to append their names to his list, and make a declaration setting forth the facts of the case, before the inspection of brokers, or the authority for such purpose, at Vegesack or Bremen-Haven.

If the captain should receive passengers aboard, other than those on the copy of the original list, and those added to it in the manner prescribed, he is liable to a fine of one hundred Bremen dollars. The same fine is imposed upon every legalized carrier of passengers, who fails to declare, upon oath, that he has taken one or more emigrants aboard, after the list was completed, without adding their names properly to it. It is enjoined upon those who may rightfully expedite vessels, to be careful not to have their passengers at the port of embarkation before their vessels are in a suitable condition to receive them. In cases of detention beyond the time agreed upon for sailing, the expense of maintaining the emigrants devolves upon the person for whom they were engaged, and if he should fail to make ample provision, he will be held responsible for any deficiency to the Bremen authorities. Ships destined for other ports than those of the United States may carry two passengers to every four tons of her burthen, United States measurement; children are to be reckoned as adults. Vessels for passengers must be sea-worthy, and in all respects well appointed.

For voyages to the United States they must have a sufficiency of provisions to supply, for a term of thirteen weeks, each passenger irrespective of age or sex, with 2½ pounds pickled beef, 1 pound pickled pork or ¾ pound bacon, 5 pounds bread, and ⅜ pound butter, weekly; to which must be added, for the passage, 34 pounds meal, beans, peas, peeled barley, rice, plums, and "saur kraut," ¾ bushel of potatoes, 1½ pounds molasses, 1½ pound coffee, ¼ pound sugar, ⅛ pound tea, 2 quarts vinegar, besides a sufficient quantity of sago, wine, and medicines for children and sick persons. Seventy-five gallons of water must be taken aboard for each passenger when the vessel sails for New Orleans or a port in Texas, and sixty gallons when she sails for any other port in the Union.

The quality, as well as the quantity of provisions, taken aboard for the voyage, must be submitted to the inspection of a person commissioned for the purpose, who must be satisfied that the provisions of the law have been complied with, and whose certificate to that effect, shall be obtained before the vessel will be allowed to start on her voyage. The owner or correspondent of a vessel, which is about to proceed on a voyage to a trans-Atlantic port, has to prove to the satisfaction of the inspection of brokers that, in case a misfortune should befall the vessel, on the way to her port of destination, by which she would be unable to continue or complete the voyage, he is prepared to refund the passage money with an additional sum of eighteen Bremen dollars to each passenger, in the way of remuneration for incidental expenses during detention. This must be done by effecting an insurance on the entire amount, at one of the insurance offices in Bremen, or with other underwriters, the policy to be approved by the inspection of brokers, (a senatorial committee,) and with them deposited for the benefit of all concerned.

Annual amount of emigration through Bremen, to the United States, for the fifteen years and a half, terminating on the 13th of June, last:

Year	Number	
1832	9,792	
1833	8,086	
1834	13,185	
1835	6,811	
1836	11,600	
1837	14,372	
1838	9,312	
1839	12,421	
1840	12,650	
1841	9,501	
1842	13,563	
1843	9,844	
1844	19,145	
1845	31,358	
1846	31,607	
1847 direct	15,990	21,911
6 months, via Quebec	5,921	
Total	235,158	

Until this year, no emigrants were forwarded to the United States from Hamburgh and Bremen via Quebec, but since the first of May, the number is that which I have respectively stated.

The number of British emigrants proceeding, via Quebec, to the United States, during the same period, I could not ascertain with any degree of accuracy, but from all the information which I collected, I am of the belief that it amounted to at least 10,000. The circumstance of so many emigrants having taken this new route, is altogether attributable to the act and its amendment, "to regulate the carriage of passengers on merchant ships," passed at the last session of Congress.

Ship-brokers, and their agents had engaged, as usual, in the winter and early spring months, emigrants, at stipulated prices for passage, for a subsequent period; but as the provisions of the acts referred to went so speedily into effect, after their approval, vessels could not be obtained to sail with passengers to a port in the United States, (at anything like the rates which they previously accepted,) after the publication of the acts. The ship-brokers, therefore, in order to secure themselves from ruinous losses, influenced the emigrants to change their destination, as originally agreed upon, for Quebec. Ships under contract, in many instances, made similar arrangements. It is scarcely necessary for me to remark that a vessel may carry a third more passengers, counting children above the age of one year as adults, to a British port in North America, than she is entitled to carry to a port in the United States.

The comparative tabular statement below, of emigration to the United States, through the ports of the United Kingdom, France, (Havre,) Belgium, Holland, and Germany, for the four years and a half, terminating on the 30th of June, will show the annual increase of embarkations for the period embraced:

Years.	United Kingdom.	France.	Belgium.	Holland.	Hamburgh.	Bremen.	Total.
1843......	28,875	8,553	2,749	1,648	1,756	9,844	53,425
1844......	49,600	16,660	2,961	2,363	1,774	19,145	92,503
1845......	58,558	23,500	4,549	4,831	2,398	31,358	125,194
1846......	82,289	32,281	11,402	5,585	3,971	31,617	167,145
1847......	101,767	16,000	8,900	5,547	3,317	15,990	168,700
Indirect...	10,000	....	....	....	1,258	5,921	
Total...	331,089	96,994	30,561	19,974	14,474	113,875	606,967

The increase of 1844 upon 1843 amounted to $73\frac{1}{7}$ per cent.; 1845 upon 1844, $35\frac{1}{3}$ per cent.; 1846 upon 1845, $33\frac{1}{2}$ per cent.; and the first half year of 1847 upon the entire year of 1846, $\frac{15}{16}$ of one per cent.

The emigration, from the beginning of January to the end of June of this year, was composed, as far as an estimate can be correctly made, of 85,000 Irish; 10,000 English; 10,000 Scotch; 61,000 Germans; 1,547 Dutch; 1,000 Swiss; and 153 French. Counting two souls, with children at the breast, to every five tons burthen of a vessel, the amount of tonnage employed in moving this emigration from Europe to the United States was 421,750, two-thirds, at the lowest estimate, of which, from statements before me, was covered by the American flag.

It is not possible to form anything like a correct estimate of the amount of money carried by the emigrants to the United States annually. In former years I knew several instances where German peasants embarked with credits, on commercial houses, for sums of $20,000. From answers, to repeated inquiries which I have made upon the subject, of persons well informed, I am induced to believe that the amount taken this year, in the aggregate, per head, was larger than usual, averaging, at a moderate calculation, $75, and making a sum total of $12,652,500.

The emigration from the ports of Europe, other than those which I have noticed, is quite inconsiderable; for the first six months of this year the number could not have exceeded 1,500.

There is but little emigration from the west and north of continental Europe to any other part of the globe except the United States. The annual embarkations from Hamburgh and Antwerp to Brazil, amount to about 1,000, and from Bremen and Hamburgh to South Australia, to an equal number.

From the lists which I was permitted to examine, at the different ports, I am enabled to state that about ninety-five hundredths of this year's emigrants are farmers, and agricultural and other laborers. But very few mechanics have embarked.

It will be perceived that I have made no mention of the emigration from the States bordering upon the Baltic. Until this year it has passed through Hamburgh; but since April, several hundred persons have embarked from Sweden and Norway, and Denmark.

Having presented to you, in as explicit a manner as I well could in a dispatch, the general purport of the enactments and decrees of Great Britain, Belgium, Holland, Hamburgh, and Bremen, respecting emigration, as well as all the information that I succeeded in procuring in connexion with this part of my mission, I might here, with apparent propriety, stop, but I should have the consciousness of not having discharged my whole duty, were I to omit the contribution of my mite for remedying the existing evils attendant upon immigration into the United States, as emigration is now conducted in one or more of the nations of Europe.

The system of Great Britain is exceedingly imperfect. It is admitted to be so; and I was assured that there was an earnest desire among statesmen and politicians of all parties to make further improvements upon it than those contained in the act of July as soon as they knew how. The subject is assuredly an embarrassing one, but this year's experience renders early actioni ndispensable.

Up to August 12 about one passenger in every six, from Great Britain, died or was dangerously ill on their passage to, or after their arrival in, Canada.

Much hesitation is manifested to adopt a measure that would reduce the number of emigrants that a ship is allowed to carry, or compel her to provision those received aboard properly, lest by raising the price of passage it may create insurmountable obstacles to many who are disposed to embark. The object of providing the one pound of bread per diem for each passenger was to guard against utter destitution. It surely never could have been contemplated that an adult should rely upon so scanty a portion for necessary sustenance from the day of sailing to that of his landing. Yet to what an immense extent has this been practised in voyages to Quebec, and occasionally in voyages to ports in the United States. The indigent Irish unfortunately avail of this pittance, and arrive out in a condition but little better than that of starvelings, so emaciated and prostrate that they have to be conveyed forthwith to hospitals. It would be better that vessels should not be required to supply the one pound of bread per diem unless they, at the same time, supply a sufficiency of other suitable food, because this serves the broker's interior agent as an efficient instrument to decoy the poor, by deluding them into a belief that they will be abundantly fed at sea.

It is difficult to determine with a tolerable degree of accuracy what number of passengers may be carried to the tonnage, or superficial mensuration between decks, in a vessel, with safety to the health of all aboard. If only one pound of bread is furnished to each, ten passengers to every hundred tons would be too many. If provided with such an allowance of food as is supplied by vessels sailing from the Weser and the Elbe, no sickness would likely be experienced where the emigrants all start entirely free from disease, from two to every four tons.

In corroboration of this view of the matter, I make the following extract from the report of the board of health of Quebec, dated August 12, 1847:

"The Larch, reported this morning from Sligo, sailed with 440 passengers, of whom 108 died on the passage and 150 were sick. The Virginius sailed with 496; 158 died on the passage, 186 were sick, and the remainder landed feeble and tottering; the captain, mates, and crew were all sick. The Black Hole of Calcutta was a mercy, compared to the hold of these vessels. Yet simultaneously, as if in reproof of those on whom the blame of all this wretchedness must fall, foreigners—Germans from Hamburgh and Bremen—are daily arriving; all healthful, robust, and cheerful."

I am, however, disposed to believe that our act of Congress, approved on the 22d of February last, as construed by the Secretary of the Treasury in his circular dated May 13, will, in the main, have a salutary tendency with respect to the space allotted in the steerage to each passenger. Fourteen superficial feet of deck, berth included, are to be appropriated to an adult, on such vessels as do not pass within the tropics; but in cases where they pass within the tropics, twenty superficial feet must be allowed to each. Under this provision a vessel proceeding to New Orleans by way of the Bahama Islands, that would be entitled to carry 200 passengers, could only take 140, if she took the favorite, because less perilous (route) around the Island of Cuba. At $30 per head a difference would, consequently, be occasioned in her receipts, from passengers, of $1,800. It is clear, to my mind, that no such result, in the workings of the law, were contemplated at the time of its enaction. From any European to any United States port, in well provisioned and well policed vessels, a space of fourteen superficial feet to a passenger is believed to be sufficient to ensure his accustomed health, if proper attention be paid to personal cleanliness. I am the more inclined to this opinion, from the facts which have fallen under my observations connected with the colonization of South Australia. Emigrants embarking at ports on the North Sea, for Adelaide, are usually out 120 days, and, although much of this time is passed within the tropics, they nevertheless generally arrive in excellent condition.

The superintendent of emigration to Australia, at Bremen, in a letter of the 11th instant, a copy of which I forward herewith, has obligingly communicated to me much valuable information. Among other things he says:

"In 1844 I commenced the agency on invitation of a member of the 'South Australian Company,' and have now completed eight expeditions of 2,000 persons altogether, and I am now organizing an expedition for November and another for next spring. I have no doubt the current will constantly and gradually increase as the accounts are encouraging; but it never will detract from America any great number, as the price of passage is necessarily double that to America, owing to distance, and also land is much dearer in Australia than in America. The British government sells land in quarterly auctions, in sections of 80 acres, at the fixed price of one pound per acre. * * * * In my expeditions I have been less restricted than the new American law restricts the transportation of passengers, and have in fact placed as many people aboard as I had room—making the berths in two rows, placing five persons together in a berth, allowing for each eighteen inches width, and having two tiers in each row. A ship of 500 tons will, with a between decks of eight feet high, admit of twelve berths in a tier × 4=240 passengers. This is ton measurement, equal to 350 lasts of Bremen. A ship of 100 tons more will not carry any more passengers. I have never had any disease on the passage, the births having always exceeded the deaths; but I have never sent away a vessel without a physician aboard, as is required by the English law."

There cannot be a question but that our laws "regulating the carriage of passengers in merchant ships," are susceptible of great improvement. I will briefly state the additions required, as they occur to my mind after much reflection:

1. That the duration of a voyage of passenger merchant ships clearing from any of the ports of Great Britain, or from the Atlantic ports of France, Spain, and Portugal, to a port in the United States, shall be computed at seventy days; that the duration of a voyage of passenger merchant ships, clearing from any of the ports on the North sea to a port in the United States, shall be computed at ninety days; that the duration of a voyage of passenger merchant ships clearing from other European ports to a port in the United States shall be computed at one hundred days. That for every passenger received aboard they shall provide a weekly supply (for the term specified, respectively,) of 5½ gallons pure fresh water, 2½ lbs. salted beef, 1¼ lbs. salted pork, or instead thereof 1 lb. side bacon, 5 lbs. bread, 2 lbs. flour, 1 lb. beans, 1 lb. peas, 1 lb. rice, ½ lb. oat-meal or peeled barley, 12 lbs. potatoes, ½ pint molasses, ⅛ lb. sugar, ⅛ lb. coffee, $\frac{1}{16}$ lb. tea, ¼ lb. butter; salt, pepper, and vinegar in proportion, 4 lbs. coal, besides all necessary cooking utensils and competent cooks. Also a box of fresh medicines, and an abundance of arrow-root, sago, castor-oil, and Epsom salts.

2. That every passenger shall be examined by a competent physician at the port at which he embarks, the day before his departure from the haven in which the vessel is moored, and pronounced free from all contagious or other distempers; and that he (or she) shall establish, to the satisfaction of the physician, that he has not been, during the six months last past, in any hospital or elsewhere under medical treatment for the typhus or starveling's fever.

3. That every passenger shall be required to bathe him or herself

carefully one or more times; to be cleanly apparelled throughout; to have all clothing composed of cotton, flax, or hemp made thoroughly clean and dry, and that manufactured from other materials well aired, as well as bedding, trunks, and boxes, before going to sea.

4. That the between decks of every passenger merchant ship shall be appropriated exclusively for the use of the passengers, none of whom shall be permitted to stow in it more clothing than is necessary for a change, from time to time, as they may require.

5. That the bilge water of every passenger ship shall be pumped out, and the hold made as clean as possible, before she receives emigrants aboard; and that she shall have wind sails down the hatchways so as to communicate, as occasion may require, a sufficiency of pure air to the between decks.

6. That the captain, or one of his officers, of every passenger ship shall assemble the passengers on deck daily at 9 o'clock, a. m., weather permitting, and call a roll of their names, and all absentees who plead sickness as a reason for non-attendance shall have suitable medicines administered to them.

7. That no passenger ship shall be permitted to disembark any portion of her passengers in a port of the United States until the captain has established, to the satisfaction of the proper authority thereat, that no person named in his list is fleeing from punishment for a crime or crimes committed, or has been discharged from a house of correction, or forwarded or turned out of an alms or poor house.

8. That passenger ships shall effect an insurance on the gross amount of the passage money which they receive, with the addition of thirty-three and a third per cent. thereon for the benefit of the passengers in the event of casualities on or in the region of the American coast, or elsewhere, if they shall prosecuute their voyage to the United States.

9. That the captain of every passenger ship shall procure, as evidence that he has complied with the obligations imposed previous to the sailing of the vessel under his command, a certificate from the authority or authorities whose business it may be to attend to such matters, that all requisitions, as far as they could transpire before the commencement of the voyage, have been scrupulously observed. Such certificate or certificates shall be duly authenticated under the consular seal of the United States at the port at which the vessel clears.

States, whose kindly sentiments for our country influence them to legislate upon the subject of emigration with an eye to the general well-being of the Union, are exceedingly discouraged when they learn that such persons as have been prevented from embarking at their ports for ourshores succeed in obtaining passages elsewhere.

It is obviously the true policy of the United States to confine, as far it can be done justly, the embarkation of emigrants to the ports of the country in which they resided. I was assured in England that the Germans passing through London, Liverpool, and Hull, on their route to the United States, are, in consequence of their utter ignorance of the English language, the constant dupes of knaves.

In conclusion, I cannot decline stating in the most positive manner, after diligent inquiry, that no disease bearing any resemblance to the fever so prevalent and fatal in Ireland, and in vessels sailing from Liverpool and Irish ports, has been discovered to exist at either of the western continental ports of Europe, or in vessels clearing from them with passengers.

I have the honor to be, sir, with faithful esteem,

Your obedient servant,

A. DUDLEY MANN.

No. 3.

Communication of Samuel Hall, of East Boston.

EAST BOSTON, *January* 5, 1854.

MY DEAR SIR: I notice in this evening's paper that you have been appointed chairman of the committee appointed by the Senate to inquire into the causes and the extent of the sickness and mortality prevailing on board of the emigrant ships. This is a very important subject, and I trust you will excuse me for addressing you, believing, however, that it will be the wish of the committee to get all the information the public are in possession of in relation to this matter. It is a well known fact that for some time past the mortality has been on the increase, and I am happy to see a move made to remedy the defect, which, in my opinion, is very easily done if Congress has nerve enough to make a suitable law and enforce it.

I will give you my opinion what the causes are, and what I am positive will be a remedy: (The main causes are only two.) The first and the greatest cause is in allowing ships to carry passengers on two decks. You will find, on examination, that the ships that have three decks and carry passengers on two of them, that disease is almost sure to follow; this is the reason that almost every packet ship has three decks and by that means can carry double the amount of passengers; here lies the greatest difficulty. The second reason is not allowing each passenger sufficient deck surface. I think the present law does not allow enough. And thirdly and lastly, to have the ships properly ventilated. I am well aware that those people that have built those three deck ships will say that the mortality is as great on the upper between decks as the lower between decks; this may be the case, but the passengers on the upper between deck have to receive all the stench from those below, as there is no other way for it to pass off except out of the hatchways, except by some small ventilators through the decks, which are, in my opinion, of but little if any use, and most of the ships do not have them.

I have been a ship and steamboat builder for the past thirty years, and know the reasons that first induced merchants to put in the third deck, nothing more nor less than to carry steerage passengers, and here is the great evil. And I did have the honor to be one of the nine supervising inspectors under the late steamboat law for a short time, but presume I was removed from my inability to perform the duties of that office. And while I held that office I was satisfied that the greatest difficulty that we had to contend with was the space to be allowed for each passenger, especially on board the steamers between New York and Chagres, and on the Pacific.

I am, with very great respect, your obedient servant,

SAMUEL HALL.

To the Hon. HAMILTON FISH.

No. 4.

Communication from John H. Griscom, M. D., of New York, to the Special Committee of the United States Senate.

NEW YORK, *January* 14, 1854.

I cheerfully comply with your suggestion, to prepare a statement of my views of the nature and causes of the diseases and mortality among emigrants *in transitu* between Europe and the United States, together with such alterations of the present laws and additions thereto, as appear to me to be indicated for the prevention of the disasters which we have recently had so deeply to lament.

No theory is better established in the science of medicine, than that certain diseases are the direct consequence of particular external circumstances. In many instances, modern investigation has reversed the views formerly entertained. Diseases which were formerly set down as the "visitation of God," over which man was supposed to have no control, and which no foresight of his could prevent, are now known to be of his own production, and either entirely preventible, or capable of great melioration.

There are some general diseases, the precise nature of whose remote causes we cannot yet discern, and of which we have no means of prevention—such as small-pox, scarlatina, measles, puerperal fever, &c.,—but even in the case of one of these, we have now an almost certain shield, and in all of them, an improved knowledge of the laws of hygiene has disarmed them of their terrors in a large degree.

Furthermore, we now know that there are certain conditions and places in which human beings are wont to live, or to which they are incidentally subject, which give them a proneness to certain diseases of specific type, but which are unknown, except under those conditions.

The laws which govern the action of external agencies upon the human body are now established, in numerous instances, upon as firm a basis as those of Copernicus or Newton, and a disregard of them will as certainly prove destructive to health and life as a suspension of the law of gravity would make "confusion worse confounded" in the material world.

Probably there never was a class of people, or any circumstances in which human beings could be placed, in which the truth of these doctrines was more clearly exhibited, than that class known as emigrants, in their transit from European ports to our own; and there is not a fact more shocking to our sensibilities, nor more disgraceful to humanity, than the condition of these people, under these circumstances.

While we have frequent expositions of the sad estate of our poorer class of city tenantry, and the public is pained with the recital of the story of their crowded and dilapidated tenements, their packed cellars and attics, their filthy yards and streets,—too little has been said of the far more horrible manner in which steerage passengers are crowded into emigrant ships, of their destitution, the filth in which they are allowed to remain, their deficiency of food and cooking, the absence of ventilation, and the too frequent disregard by owners and masters, of the spirit, and even of the letter, of the laws, both of our own and other countries; the result of which has been an amount of disease and mortality unprecedented in modern times, under any similar circumstances.

There are three diseases which especially conduce to these results, viz: *typhus fever*, *cholera*, and *small-pox*. Of these three, that to which the emigrant is most prone, is typhus fever, in that form commonly known as *ship fever*.

The extraordinary prevalence of this disease at the present time, and for the past half century, but especially for the past seven or eight years, is an astounding phenomenon, particularly when it is remembered, that we live in the midst of all the light necessary for its prevention.

My first practical cognizance of the horrible condition in which emigrants are frequently found on shipboard was in 1847, when, as a member of a committee of the New York Academy of Medicine, I visited the quarantine establishment to enquire into the medical history of the typhus fever, then extensively prevailing, and crowding that institution with patients. On that occasion we visited the ship *Ceylon*, from Liverpool, which had come to anchor a few hours before, with a large cargo of passengers. A considerable number had died upon the voyage, and one hundred and fifteen were then ill with the fever, and were preparing to be removed to the hospital. Before any had yet left the ship, we passed through the steerage, making a more or less minute examination of the place and its inhabitants; but the indescribable filth, the emaciated, half-nude figures, many with the petechial erupture disfiguring their faces, crouching in the bunks, or strewed over the decks, and cumbering the gangways; broken utensils, and dêbris of food spread recklessly about, presented a picture of which neither pen nor pencil can convey a full idea. *Some were just rising from their berths for the first time since leaving Liverpool*, having been *suffered* to lie there all the voyage, wallowing in their own filth. It was no wonder to us that with such total neglect of sanitary supervision, and an entire absence of ventilation, so many of such wretched beings had perished, or were then ill of fever; it was only surprising that so many had escaped.

Shocking as this case was, it has been frequently surpassed, at least as far as figures are concerned. In 1842 the ship *Eutaw* gave one hundred and twenty to the hospital on arrival; in 1837 the *Ann Hall* sent in one hundred and fifty-eight; while as far back as the year 1802, one hundred and eighty-eight were taken from the *Flora*, two hundred and twenty from the *Nancy*, and two hundred and fifty-nine from the *Pene-*

lope. In 1851 the number of deaths at sea, between Liverpool and New York, rose to the astounding number of 1879, almost wholly the result of ship fever.

In addition to this the poisonous influence which has become infused into those who have escaped death or sickness on ship board, lies dormant for a few days or weeks after debarkation, and sooner or later developes itself, and brings many of them to the hospital, where from fifteen to twenty per cent. more are added to the list of dead. Thus there were treated in the marine hospital, on Staten Island, in 1852, 3,040 cases of ship fever, of whom seventeen per cent. died. These were all emigrants; and we must add to these the cases of the same disease, of the same people, which were treated in the large hospitals at Flatbush, Ward's Island, and Bellevue, at the city hospital, and at other places throughout this State and the States immediately adjoining, nearly all of whom arrived at the port of New York alone.

In considering the hygienic aspects of emigration, we start then with the remarkable fact, that of those who embark for an Atlantic voyage, on any one of a certain class of ships, *one in every twelve* of them but steps into a coffin; nearly nine per cent. will either never reach the promised land, or will die soon after.

The general causes, as well as the means of prevention, of this disease, are so plain, as not to require a medical education for their comprehension, but may be made clear to ordinary intelligence, and the vast importance of the subject will justify an allusion to both in the present communication.

Ship fever, as it is termed from the places of its greatest prevalence, is the product of a *miasm*, as distinct as that of marshes, which produces intermittent fever; and this miasm is itself as necessary a result of certain prior circumstances, as the marsh miasm is the product of marshes. And further, the means for its prevention are as clear and controllable in the one case as in the other. Thus if an offensive marsh be thoroughly drained and dried, its peculiar miasm, and the disease which it caused, will disappear; and so by preventing the formation of the miasm of ship fever (as easy of accomplishment as the other) that disease will in like manner be prevented, or avoided.

What then are the circumstances which give rise to this miasm of typhus? There are certain essentials to its creation, which I will enumerate in the order of their importance, beginning with the least:

1st. The confinement of numbers of people together, in apartments disproportioned in size to their requirements of wholesome respiration.

2d. The retention in the same apartment of the excretions from the bodies of the individuals thus confined; such as the matter of perspiration, the carbonic acid gas and moisture from the breath, and other more offensive excretions. These acted on by the artificial heat of the apartment, or even by the natural heat of the bodies alone, will become decomposed, and produce an effluvium which will react poisonously on the persons living in it.

3d. Too great exclusion of pure air.

As to the first of these causes, the number of persons, and the size of the apartment necessary to produce this miasm, are merely relative. A *single individual,* subjected to these circumstances a sufficient length of time, may excite typhus fever in his own person. By shutting one up in a small room, at the same time excluding the external air, and allowing the excretions of his body to accumulate around him, typhus fever will almost certainly appear, sooner or later, according to his constitutional power of resisting the poison. On the other hand, an apartment may be ever so crowded, without the least danger from this source, provided that, from the first, *ventilation and cleanliness be thoroughly and constantly maintained.*

In support of these views it were easy to quote abundantly from the most authoritative medical writers; but I deem it unnecessary, considering them to be established beyond dispute, and not wishing to lengthen this communication unnecessarily.

With this brief explanation of the general causes of typhus, the reasons of its prevalence in the steerages of passenger ships are very apparent. In great numbers of them, *all* the circumstances enumerated above, as necessary for the creation of this disorder, are found in active operation. But to account for the extraordinary amount of it during the past few years, some other and more particular cause seems necessary. Steerages crowded with careless passengers, and badly ventilated or not ventilated at all, have been known, before the period alluded to, to escape its devastations, and it is therefore reasonable to infer the existence of some *specific* cause, in addition to the general ones which have been mentioned.

We find ship fever, within a few years, to have prevailed most frequently and extensively in those vessels which ply between several ports of Great Britain and this country, and this fact, together with an examination of the passengers, points unerringly to the *famine* which desolated a large section of that kingdom, as the additional cause alluded to.

This, however, was not a direct, but only an *indirect* or *remote* cause of the evil. The long privation of food before embarkation, so effectually reduced both bodily and mental strength, that when once on board and in a berth, the most powerful inducements were insufficient to get the passengers out and on deck. Hunger itself would fail to bring them forth. This very indifference to self-exertion, this irresistible prostration and indolence, if unchecked by the police of the ship, results in the rapid accumulation of filth of every possible description, and the speedy generation of the typhus miasm. Starvation alone, however, never yet produced typhus, or any other disease; but, in the way pointed out, it is a most potent predisposing cause, and facilitates the action of the other necessary circumstances.

In connexion with this branch of the subject, another source of its developement on board ship demands notice. In the cabins and hovels —the homes—of these famine-stricken people, typhus fever raged a long time, and doubtless prevails extensively yet, produced by the same general and specific causes as have been described. The emigrants leave for the seaboard, and straightway enter into the ships, unpurified and unwashed, reeking with the fever miasm of their dwellings. Into the crowded and confined steerage they precipitate themselves for rest and escape from starvation and death. But they have brought the enemy with them, though unconsciously; the fatal seeds are but sown in a fresh soil, and, as if in a hot-bed, they spring up with increased vigor, infecting not only those who introduced them, but others who have heretofore been free from the infection. *One such case* on board a crowded ship, especially if badly ventilated, *may cause the death of numbers.*

Against this source of infection the English government has attempted a preventive measure, with what success will be hereafter alluded to.

A very important consequence of this overwhelming prostration of strength is, that the food with which these people are supplied on board ship, even if sufficient in quantity, (which is not always,) is very often so badly cooked as to operate upon them injuriously rather than otherwise. So great is the difficulty often, among from 300 to 1,000 people, of finding a proper time and opportunity for cooking, that it is a common occurence for them to swallow their flour or meal only half-cooked, or even simply mixed with warm water, if indeed *warm* water can be had. The effect of this kind of diet is but to add other evils, such as dysentery and diarrhœa to the typhus miasm, with which the steerage has become infected, rendering it more virulent, and the debilitated inmates more susceptible to its influence, while a well-fed person will longer and more effectually resist it.

For the prevention as well as the cure of typhus, it is necessary that the physical stamina be well maintained by appropriate food in sufficient quantity. With ordinary strength of body and elasticity of spirit, few persons can be induced to remain below deck many hours together; and while the pure air of the ocean directly increases animal vigor, it is also the surest preventive of typhus. Even the half-starved emigrant would find his energy and spirits revive, if compelled by a rigid sanitary police to make frequent visits to the ship's deck.

Famine, therefore, though a frequent precedent, and a powerful adjuvant, is only an indirect cause of the fever as we find it on ship board and in our hospitals; but thus we must continue to be burdened with it, as long as poverty-stricken emigrants are admitted into the transport ships in such great numbers, with food so insufficient in quantity and quality, and with such total absence of sanitary police during the voyage.

From what has been said, it will be readily inferred that in the prevention of typhus fever pure air possesses great value. Too much reliance cannot be placed upon it, either for this purpose, or for subduing the intensity or arresting the progress of the disease. Of its efficacy as a remedial agent, a striking instance, among many others that might be mentioned, occurred at the New York Quarantine Hospital, under my immediate notice, during my connection with the State Emigrant Commission. A new building was erected on the summit of a hill within the enclosure, into which some forty patients were conveyed from the other over-crowded buildings. Although these had been kept in as good condition as possible as respects both cleanliness and ventilation, though there was no specific mode for the latter, yet the influence of the fresh atmosphere of the new building upon these patients was most decided and immediate; a load seemed to be lifted off them, and several who it was feared would die began at once to improve, and rapidly recovered. A more remarkable proof still of the curative powers of pure air occurred in 1837, at Perth Amboy, the details of which will be found in a paper published in the "Transactions of the New York Academy of Medicine," and which is appended hereto. (See page —.

It has been a matter of conjecture with some, why typhus fever and many other infectious disorders are more abundant and fatal in cold weather than in warm; its increase has been attributed by some to the low temperature. But the true reason undoubtedly is, that in winter the external atmosphere is more completely excluded from our dwellings and hospitals by the closing of doors and windows, which in warm weather are open, and freely permit the ingress and egress of air. Hence in winter the greater necessity of *artificial ventilation*. The same reasoning applies to passenger ships in cold or stormy weather, when the hatches are kept closed; artificial ventilation, necessary at all times, is then more abundantly demanded.

Persistence of the Contagion.

There is another fact connected with ships as well as with hospitals and dwellings, which has a very important bearing on this subject. The miasm which has been spoken of, has the property of attaching itself to clothing, bedding, furniture, and to the walls, ceilings, and floors of apartments; it is absorbed by them, and adheres with considerable tenacity, whence it is ever ready, unless thoroughly destroyed and removed by cleansing and the use of disinfectants, to issue forth, and like the leaven hid in the meal, to leaven the whole atmosphere of the apartment with its poisonous influence. Into a room in which a case of typhus fever has once existed, even for a short time, it is unsafe to enter, unless the room, and everything in it, has been first subjected to a thorough airing and purification. Here then is a constant source of danger, and which will probably account for many instances of its devastation on ship-board. A vessel in which this disease has once occurred will have the miasmatic poison clinging to its sides, ceilings, and floors, whence it is impossible to eradicate it without the most thorough airing, ablution, and disinfection, such as, I presume, no vessel engaged in the European passenger trade has ever yet received. In hospitals and dwellings, with hard finished walls and painted wood work, this fact is often demonstrated;* but in the peculiar structure of a vessel's inner walls, without plaster, paint, or whitewash, with thousands of crevices and cracks inaccessible to the scrubbing brush or any other purifying implement, without windows for the free circulation of air, we see the *perfection* of a place for the long retention of the poison, and for its propagation for months after, when the steerage shall be again crowded with human food for it to fasten and grow upon.

There is abundant authority for the substantive, adhesive, and propagatory character of this infectious principle. Dr. Rush says this poison may remain in fomites *six months* without losing its active properties. Dr. Nathan Smith presents evidence to the same effect, and M. Gendron, a French physician, states some facts which seem to show

---

* See a case reported in the transactions of the New York Academy of Medicine hereunto appended. (See page —.)

that the contagious matter of the disease may remain active in a bed for two or three years.

Dr. Millan (report on the diseases of London) remarks that the houses of the poor in London are often so little taken care of, that in the apartments where contagious fevers have existed, enough of the contagion remains to infect all the inmates who successively occupy the same premises; and he mentions some particular houses in which the fomites of fever were thus preserved *for a series of years*.

Further than this, it is believed that this poison *increases in power* by a long retention in a room, or in clothing. Thus, Dr. Wilson Phillip, in his standard and elaborate work on fevers, in his chapter on typhus, says: "Fomites often retain contagion for a great length of time, and may convey it any distance. It is a general opinion, that fomites more readily communicate the disease, and communicate it in a *worse form*, than the sick themselves." The celebrated Dr. Cullen makes a similar remark, and in addition states that "the effluvia constantly arising from the living human body, if long retained in the same place without being diffused in the atmosphere, acquires a *singular* virulence."

Dr. Parr, in his medical dictionary,* says: "Fevers caught by *recent infection* are *mild* compared with those which arise from *contagion long pent up, styled fomites*."

ONCE INFECTED, ALWAYS INFECTED, UNTIL DISINFECTED, should therefore be said of every vessel; and it is for the non-application of the last clause of the sentence, doubtless, that we are in many instances pained with the reports of death's doings on shipboard, when the remote causes, before alluded to, do not exist as in 1847, and thereabouts. It is thus they are first made into, and thus they are continued to be, floating "Black Holes," scarcely surpassed by the great original at Calcutta.

Moreover, the bunks or berths on these vessels are generally constructed of the cheapest kind of boards, often in the rough state, put together without any nicety, and their whole arrangement of the most temporary character. Nothing of the kind could be better adapted for harboring and retaining the fever miasm. At the end of the voyage they are sometimes without disinfection, or even washing, taken down, and with all their filth and miasm adhering to them, are stowed away either as dunnage, amid the return cargo, or in bulk, and appropriated to their original purpose on the next hitherward voyage. Now it is evident that the next cargo of emigrants which such a vessel receives may be composed of perfectly healthy people; and though they may be well supplied with stores, and the bedding, clothing, persons, and habits, be of the cleanest and best character, yet in the *ship itself* are the seeds of disease which, night and day, they must be in close contact with, and which, under ordinary circumstances, will, in all probability, attack some of them. The pestilential wave, once started, will roll on to engulph one after another of the entrapped and helpless victims who have confided too readily in the purity of the vessel and their own cleanliness and care.

But even without this additional source of danger, concealed in the temporary structures alluded to, and supposing all the old lumber to be

* Cited by Prof. J. M. Smith.

destroyed, and new substituted each voyage, still the permanent timbers of the vessel form a nidus for the poison of the surest kind, whence it may issue and display its powers upon the fresh victims.

Cholera.

Our attention has been recently arrested by the appearance of another disorder among Atlantic emigrants, which, with swifter foot and deadlier aim than typhus, has decimated them in many vessels. Cholera is a new feature in emigration, as it was in its epidemic form in 1832, an entirely new disease on the western hemisphere.

A native of Asia, it spread over Europe, and then swept the American continent, appearing at different intervals since, but until within a very recent period confining its devastations to the land. It has now apparently abandoned the erratic character which it assumed at the commencement, and taken up a permanent abode among the habitations of men. It even tracks him to the water's edge, and enters with him into the ship, concealing its presence until fairly launched upon the deep, beyond the reach of succor, and then, springing forth from its lurking places with resistless fury, sweeps overboard victims enough to have appeased the demands of the most sanguinary god that mythology ever invented.

We have now to inquire whether in its nature, its causes, or its progress, it bears any resemblance to ship fever, and whether we have any knowledge of the means of its prevention, or, if started in life, we may restrict or arrest its progress, as we know we may do with typhus.

In the commencement of this inquiry, we meet with one striking difference between them. Cholera, as it first appeared among us, and is now occasionally observed, is an *epidemic* disease, which ship fever never is. Of the latter, the open air is as complete an extinguisher as water is of fire, while the former has been known to move onward over whole communities and countries, heedless of seasons or temperatures, and laughing to scorn all human efforts for its arrest. Its terrific energies seemed to be exhausted only by time. It did not formerly visit alone the abodes of wretchedness, the cellars, the crowded courts and wynds, and mud hovels—though the inhabitants of these suffered in much larger proportion—but, like the hurricane, it fell upon both rich and poor. In time it passed away, and the world fondly hoped that, as it was a new visitant, its first visit was its last. It came a second time, however, but with modified severity, and although its origin was yet entirely unknown, the hope of its final disappearance still lingered. But in this we have been doomed to disappointment. True it has not again presented itself as a general epidemic, but we have had it devastating particular localities; we see it here and there sprouting up under certain conditions and circumstances, and giving too strong evidence of its being now a permanent addition to the already too long catalogue of the ills which flesh is heir to.

We have therefore had more frequent and ampler opportunities for the examination of its characteristics. What, then, let us enquire, are the localities in which it is most frequently found? What the circumstances under which it is most apt to appear? And who the people it most affects?

We answer that its localities are peculiar, its favoring circumstances easily understood, and its victims almost exclusively of a certain class. These are all *identical with those which are known to be especially prone to typhus fever.*

There is now no longer any room for doubt, that this disease is very frequently produced, and is always aggravated, by filth and foul air. Numerous instances are recorded of its spontaneous appearance in places occupied by people of filthy habits, crowded, badly fed, and ill ventilated. It cannot be accounted for in these localities, on the same hypothesis as typhus fever, that is, by the reaction upon the system of a poisonous miasm, which must first be produced, though it may be upon another theory equally sound. The typhus miasm is created by the decomposition of pent up animal secretions, is of a peculiar character, and can cause no other disease than typhus; yet it is, I think, demonstrable, that cholera is the result of the action upon the human system of foul air, defective nourishment, and other vitiated circumstances, similar to those from which typhus eventually springs. Their *modus operandi* in the causation of cholera is probably this, that a more or less protracted residence amid these depressing circumstances, results in a highly vitiated state of the system, a peculiar *cachexia*, from which, under a favoring condition of the general atmosphere, the new disease is ushered into life, ere a sufficient time has elaped for the creation of the typhus miasm; as if the formation of this poison was a too tardy process, and Providence deemed it necessary to send a swifter winged messenger of death, in these latter days, to punish his creatures for their continued violation of the laws given for the preservation of their health and lives.

The correctness of this view of its causation, and the relation of these subjects to each other, as cause and effect, as set forth recently by a distinguished member of the British cabinet, will undoubtedly receive the unanimous assent of the medical profession:

"The Maker of the Universe has established certain laws of nature for the planet in which we live, and the weal or woe of mankind depends upon the observance or the neglect of those laws. One of those laws connects health with the absence of those gaseous exhalations which proceed from overcrowded human beings, or from decomposing substances, whether animal or vegetable; and those same laws render sickness the almost inevitable consequence of exposure to those noxious influences. But it has at the same time pleased Providence to place it within the power of man to make such arrangements as will prevent or disperse such exhalations, so as to render them harmless, and it is the duty of man to attend to those laws of nature, and to exert the faculties which Providence has thus given to man for his welfare.

"The recent visitation of cholera, which has for the moment been mercifully checked, is an awful warning given to the people of this realm that they have too much neglected their duty in this respect, and that those persons with whom it rested to purify towns and cities, and to prevent or to remove the causes of disease, have not been sufficiently active in regard to such matters. Lord Palmerston would, therefore, suggest that the best course which the people of this country can pursue to deserve that the further progress of the cholera should be stayed, will be, to employ the interval that will elapse between the present time and the beginning of next spring, in planning and executing measures by which those portions of their towns and cities which are inhabited by the poorest classes, and which, from the nature of things, must most need purification and improvement, may be freed from those causes and sources of contagion, which, if allowed to remain, will infallibly breed pestilence, and be fruitful in death, in spite of all the prayers and fastings of an united but inactive nation. When man has done his utmost for his own safety, then is the time to invoke the blessing of Heaven to give effect to his exertions."

But, on the other hand, granting that it cannot thus be generated *de novo*, and that its endemic or sporadic appearance must be the result of the germinature of seeds which were sown by some former epidemic, (as some appear to hold,) there can yet be no reasonable doubt, that certain soils are more favorable for their germinature, and that the most favorable are those ærial soils, so richly manured by concentrated human effluvia and filth. Without such provocatives and stimuli to growth, these suppositious seeds would lie dormant, for aught that we are aware, perhaps as long as the cereal grains found with the mummied bodies of the pyramids.

Again we may be told that there have occurred instances in which neither of these theories of causation is sufficient to account for all the circumstances under which it has appeared. Thus it has been observed to make its appearance on some vessels in a very sudden manner, and, after a few days continuance, to have as suddenly ceased and disappeared. And there has been generally noticed, in these cases, either a decided change of wind, or some other alteration in the state of the weather, while its visitations have also been said to be confined to certain latitudes.

I do not see in these cases sufficient to charge the theory advanced with insufficiency or inconsistency. In some instances it would indeed seem as if the state of the weather exerted a decided influence both upon the origination and upon the duration of the attack; but in some other instances it would appear to have been independent of direct influence from that quarter. In one case we are told the disease attacked the passengers before the ship left her moorings in the river Mersey and prevailed so severely as to induce the master to discharge the whole company and take in another of a better class. And though we may admit that there is in the constitution of the general atmosphere something of an occult or mysterious character, which will produce this strange disease, yet the opinion cannot be dismissed or disregarded, that there must be in the *local* atmosphere powerful predisposing and supporting causes, otherwise it would not prevail so uniformly among people of a particular class and character. Separated as they are only by so narrow a line as that which divides the cabins from the steerage, on one side we find total exemption, and on the other great proneness to the disease. It enters the ship by the hatchways, and not down the cabin stairs.

The same rules which are required to be observed for the prevention of typhus, are therefore demanded for the avoidance of cholera, with perhaps the single exception of that which refers to the destruction of the contagious miasm which typhus leaves behind it, and with which it impregnates the furniture and timbers of the vessel. There is no evidence of any similar result from the prevalence of cholera, though after the existence of such a disease, in such a place as a ship's hold, it should not be deemed fit to be entered by others, much less suitable for a crowd of emigrants, without a most thorough cleansing and disinfection.

Another theory by which the existence of cholera on ship board is accounted for, in the minds of some, is, that it is brought on board by the emigrants themselves, from infected districts, and sooner or later breaks out and spreads. This idea presupposes the existence of a cholera virus or miasm, which may hang about a person's clothing or body, and lie dormant there for several days. Of this there does not appear sufficient proof, though the opinion is entertained by many, that the disease has given some evidence of a contagious character. But admitting it to be true, still the virus, like that of typhus, must have a favorable soil to sprout and grow in, and that is furnished in perfection on board an ordinary emigrant ship; and that it does not infect the cabin shows that the steerage affords peculiar facilities for its growth.

The same remarks will apply to the theory which has been advanced, of there being zones of atmosphere of a choleraic tendency, which is exerted upon the vessel as it passes through them.

Small Pox.

With regard to small pox, the third in rank of the diseases which have afflicted emigrants, its nature and its means of prevention are too well known to require anything more than a single passing recommendation.

I believe that this disease appears chiefly on vessels from French or German ports; and it is due to the passengers thence, as well as to ourselves, that some understanding should be had between the respective governments, in reference to the rigid inspection of passengers, and the enforcement of vaccination before embarkation, and the purification of the vessels in which the disease has occurred, after their arrival here.

Remarks on the law of 1848, and proposed amendments.

Section 2. The arrangements for ventilation, as required by this section, are exceedingly deficient, both in amount and power. Only one receiving tube, of twelve inches in diameter, is required for the supply of air for *two hundred persons.* Even under the influence of a good breeze, this would be insufficient for the healthful change of the atmosphere of a steerage containing one quarter that number of people, though they should be of the best character for cleanliness of person and habits. Larger tubes, and more of them, are necessary.

Another defective point is, that these tubes are not required to terminate below in such a manner as to distribute the air in different directions through the apartment. A single current is produced by them, which cannot be expected to permeate into the numerous recesses, nor to disperse itself over or between the berths.

The *exhausting* tubes, as they are generally made, (without an exception, as far as I have seen,) are *not* exhausting tubes. They will rather prevent than facilitate an upward current through them. The form of cowl, capable of producing the greatest degree of exhaustion by the action of the wind upon it, has been a matter of much philosophic speculation, and several have been invented, each of which claims superiority, and all are effective in a greater or less degree. But it is confidently affirmed that the form usually found on ships rather than produce, will more often prevent an upward current.

It is, moreover, a question whether an exhausting cowl, even of the most approved pattern, is necessary. The greatest amount of foul air which one could bring up from below would probably afford a very insignificant amount of relief under the most favorable circumstances. I would recommend that there be no exhausting cowls or tubes, as such, but that all should be used as receiving tubes, and the exhaustion be looked for through the hatches, and that the terminus of each receiving tube be arranged so as to distribute the air more widely. "Plank shear" ventilators should be required in every vessel hereafter built, and also in those into which they can now be introduced. These constitute exhaust openings of much efficiency and continuance.

Another serious objection to this section is its "proviso." It submits the whole matter of ventilation to the judgment of an inspector of the customs. In this there is no guaranty that even the letter of the law, defective as it is in this respect, will be executed, and less that its benevolent spirit will be. The whole question of means of ventilation is thus left to the decision, virtually, of—*we do not know who*—and most probably some one ignorant of its first principles. The means should be, as far as practicable, strictly defined by law, and what cannot be thus defined submitted to the discretion of an officer thoroughly informed on the subject.

Section 4. This section, in not requiring cooks for the passengers, is inferior to the English law, which does require them. It is more important, for obvious reasons, that the United States law should require them, and also that the cooking be done *for* the passengers, and not *by* them. Fuel for cooking is required to be distributed to the passengers weekly, which pre-supposes the cooking shall be done by the passengers, each for himself—a thing manifestly impracticable for from two hundred to nine hundred to do at one fire, and improper if it were practicable. That portion of the section should be repealed, and substituted by a requisition for the appointment of a sufficient number of cooks as a part of the ship's complement, whose exclusive duty it should be to cook for the passengers the food furnished by the ship, as well as any articles furnished by the passengers themselves, under such regulations as may be necessary for the preservation of order and discipline.

To a deficient quantity and quality of food is attributable a great proportion of the sickness and other evils of the steerage. Good meal, in sufficient quantity for the demands of appetite, not only does not, if but half cooked, supply nourishment and strength, by reason of its not being assimilated, but it acts as an irritant upon the intestinal canal, producing diarrhœa or dysentery, debility and prostration, and renders the recipient an easier prey to the miasm, which he thus helps more rapidly to create. So far as the prevention of diseases generally found on shipboard is concerned, there is really more necessity for a sufficiency of substantial, well-cooked food in the steerage than in the cabin. The better housed and conditioned inmates of the latter would be far less liable to sickness than their neighbors in the steerage, under a short allowance of food, and chiefly for the reason that they have a better atmosphere, and less concentration of the emanations from their own bodies, which the absence of food will not aggravate as it does in the steerage. It is not meant that the steerage passengers require the epicurean cookery, delicacies, and condiments of the cabin; these would not add to their strength and healthfulness; but true nourishment, substantial, digestible food, is needed for the effectual support of the inmates of the steerage, under their more debilitating external circumstances, far more than for the better clothed and better lodged occupants of the cabin, albeit these are accustomed, in general, to a more stimulating course of life.

There should be furnished to the steerage passengers also, at least three times a week, a fair proportion of fresh animal food, either in the shape of meat or soup. The law requires now but ten pounds of salted pork for each passenger for the whole voyage. This is insufficient for the purposes of health. It is true that many passengers furnish themselves with ham and other forms of animal food, but a great many depend wholly upon the ships' stores; and these are they, in general, who require good nutriment more than the others.

Sec. 8. By this section passengers are permitted to be carried in an apartment *which may be less than five feet in height*, which is less than the average height of adult human beings.

I know not whether any vessel with steerage of that dimension only has ever been engaged in the transportation of *voluntary* passengers, but certainly the country is disgraced by such a permission in its laws. To say nothing of the physical oppression to which every man and woman must be subject in such a place, the pathological oppression must be inexpressible. The tendency to disease increases rapidly as the space is diminished; and the increased number of superficial feet required for each passenger is by no means a compensation for the diminished altitude, and for this reason: The enanations from the bodies and lungs at first necessarily ascends to the uppermost portion of the space occupied, and, when the ceiling is far above the heads, will pass out of reach of inhalation; but the nearer the ceiling to the head, the greater the danger of these effete matters being inspired and absorbed into the system. In a crowded room, whose ceiling is even as much as two or three feet above the head, the stratum of foul air will soon reach so low as to be inhaled. When the ceiling is but seven feet, a few inches above the head, *a few minutes* will suffice to bring the foul air within reach of the lungs; if it be but six feet, the heads of the people of ordinary height must be *immediately* immersed in their own foul gases. But at five feet, men, women, and children must inevitably and continually be steeped, head and shoulders, in a rapidly-accumulating mass of

5

corruption, which no ventilation can remove, and the horrors of which can be surpassed by nothing in history but those of the middle passage of slave ships. It were as impossible to maintain human health in such a concentration of foul gases as must necessarily arise there, even with double the number of superficial feet allowed to each passenger, as in the Grotto del Cane, while the introduction of a case of contagious or infectious disease would light a flame which could be extinguished only by a complete evacuation of the place. The possibility of such a hold being used for the stowage of passengers should be at once, forever, and totally prohibited. Even six feet height of steerage, which is the minimum limit of the British law, (§ 70,) is entirely too small for such a purpose, for the reasons stated, which cannot be controverted or avoided.

What, then, would be a minimum limit to make a steerage fit for the carriage of passengers? In answer to this question the impression will, at first be, that the ceiling should be at least high enough to permit the tallest man to walk erect in any part. And beyond that, on the principle of the *higher the better*, the greatest possible height that the construction will admit of should be required by law of all vessels intended for the passenger trade.

In connexion with this branch of the subject, another point claims serious attention. In many of the larger ships, which have three decks, there are two steerages, *one below the other*. Now all the evils which have been considered in this communication, in connexion with emigrant vessels, are incidental to the uppermost apartment, which is directly beneath the main deck, more readily accessible to the air and light through the hatches and side lights, which latter, in fair weather, are always open, making it a comparatively cheerful place. Its proximity to the upper deck also renders the latter much less difficult of access, and consequently the passengers may more easily and frequently be induced to visit the open air.

But in the lower steerage, the evils of the upper one are doubled in intensity. In consequence of its depth, its only light is from the hatches, which is mostly diffused into the upper apartment before it reaches below. The same openings which supply air to the upper, and which are inadequate for *its* full measure, are also depended upon for the supply of air to the lower steerage, and, as in the case of the light, the upper receives nearly all the benefit, small as it is. In fact, every sanitary arrangement which may be put in force on board the ship, is deprived of half its efficacy in the lower steerage, as compared with the upper.

In broad sun-light, with the hatches all open, and the vessels lying quietly at the wharf, on a recent visit to several of these three-deckers, which had arrived with large consignments of emigrants, the dirt and muck on which we trod could be felt but not seen. At sea, when lights are not permitted below, and there are many causes to intercept the few rays of daylight which struggle to descend, the condition of things cannot be seen, and *cleaning is impossible*. Almost perpetual night reigns in these sub-aqueous abodes. A residence there of thirty hours must be enough to sadden and depress the coarsest sensibilities of its inmates, and produce great proneness to disease, independently of the filth and foul air which envelopes them before they lose sight of their native shores; but at the end of thirty days, the scene is one which humanity shudders to dwell upon.

When, therefore, we consider the impracticability of preserving cleanliness and purity at such a depth from the surface, it becomes a serious question whether the second hold should not be prohibted as a receptacle for passengers.

There is another consideration which adds great weight to the argument for such a prohibition. Not only does the effluvium from the bodies and lungs of the inmates of the lower hold, more rapidly accumulate there in consequence of the greater inaccessibility of air, and the difficulty, on account of darkness, of removing the filth; but these emanations, more or less, rapidly find their way into the upper steerage, where they add to the mass of *its* festering contents, and render its ventilation more and more necessary. It is well known that typhus fever in one apartment of a dwelling, unless it be well ventilated into the open air, will poison the atmosphere, and produce the same disease in another apartment on the same floor. But far more readily will the miasm ascend to an apartment above, especially when the communication between them is always open, as in the case under consideration.

For these reasons, there can be no hesitation in the recommendation to prohibit the use of the third deck for the carriage of passengers.

Of the space proper to be allotted to each passenger.

From my remarks on the causes of typhus fever, this will readily be inferred to be one of the most important points of the whole subject under consideration, especially as in it is involved the question of the number of passengers each vessel should be allowed to carry.

Two methods of graduating the number of passengers, have been tried in the laws on this subject. The first method was by the *tonnage of the vessel.* Under this mode of computation serious evils arose, and it was abandoned. It was manifestly an improper, or at best a very uncertain, mode of defining the proper quantum of space for the passengers, as it was putting them upon almost the same footing, in this respect, as the inanimate cargo, and there were too many masters prone to exercise more care in the stowage of the latter, than of the former.

Next was devised the method of allotting a certain number of *superficial feet of deck* to each passenger. This was an improvement, and, in a majority of vessels, considerably reduced the number of passengers allowed to each, though it is said in some instances to have increased it. But it is a more accurate and tangible method of computation, and hence was an advance in the right direction, though it must yet be regarded as failing in the true principle upon which the calculation should be made.

To one who should undertake the investigation of this subject for the first time, it might seem that the object of the law was to obtain for the passenger merely standing or walking room, and sleeping space, and that the allotment of *air* was a matter of secondary consideration, or rather no consideration at all. This latter might be *inferred* as the object of the law, but its numerous omissions would make the inference a very doubtful one.

The true sanitary principle by which this matter should be regulated, is not the number of tons burden of the vessel; in other words, the amount of dry goods, or coal, or iron she can carry, nor the number of superficial feet of her decks, but *the number of cubic feet of air of the apartments allotted to passengers.*

This is the true standard of sanitary capacity, when considered independently of its means of ventilation; and this, in my judgment, should be the basis of a restrictive law, in the allotment of numbers.

In the first place, the minimum height of ceiling should be defined, and then a certain number of cubic feet of free space be apportioned to each passenger. By *free space* is meant the area of the apartment not occupied by the masts, bulkheads, casks, chains, beams, &c., all of which diminish the amount of air in the steerage. The luggage, and even the bodies of the passengers, occupy a certain cubic space, to the exclusion of an equal bulk of air, and should be regarded in the calculation.

Having settled this as the principle upon which the capacity of a vessel should be estimated, the next point of inquiry will be the extent of cubic area which each passenger should be allowed. Several elements are found pertaining to this calculation. On the principle before laid down, that *the more active the ventilation of an apartment, the more may it be crowded with impunity*, it will at once be seen that, in the allotment of space for passengers, the height of, and the freedom of circulation of air through the steerage, should be taken into account. The application of this principle would have a salutary effect upon the owners of passenger ships, in inducing them to make their steerages of the greatest possible height, and introducing the most effective means of ventilation, as thereby they would obtain the privilege of a greater number of passengers.

But, as in the case of the height of the ceiling, a minimum amount of free cubic space should be fixed by law. In relation to the question of what this should be, a diversity of opinion exists among well-informed men. It involves the question of the amount of air requisite for the wholesome respiration of an individual for a given space of time. The estimates of the amount required for an adult *per minute*, and which, at the end of that time, must be entirely removed, to avoid the risk of reinhalation, vary from four to ten cubic feet. This is with the air at rest, and not communicating with the general atmosphere. With a free communication between the general atmosphere, and that of the apartment, whereby the carbonic acid gas, and other exhalations can be freely diffused abroad, ten cubic feet per minute is probably more than is needed, but the lower figure given above is, on the other hand, too small. The famous black hole of Calcutta, we are informed, was about eighteen feet square, and it was probably not over ten feet high. This gives it an area of three thousand two hundred and forty cubic feet. On the fatal night which has given it so great notoriety, it was made to contain one hundred and forty-six persons, thus allowing to each only twenty-two cubic feet—though as each body excluded an equal bulk of air, it was probably not near so much as that. At the lowest estimate mentioned above, (four feet,) before five minutes elapsed, after the door was shut upon them, the hapless victims began to reinhale their own exhalations, and this process was, of course, repeated at each similar successive period until death began to reduce their number. They were confined there about ten hours, and although there was an open window on one side, twenty-three only survived till morning, and they were in a "high putrid fever," (typhus.)

I allude to this oft quoted case chiefly for the purpose of showing those not familiar with this topic, and who may be sufficiently interested in this subject to read this communication, what are the true principles upon which a calculation for the allotment of space for emigrant passengers should be based. Any more minute detail, in this already too long paper, of the elements of this calculation would probably be considered burdensome. I may state, in concluding this part of the subject, that a recent examination of the two steerages of one of the largest packets belonging to this port, (authorized by the present law to carry over nine hundred,) gave as the cubic space for each passenger, not deducting the room occupied by the necessary solid contents, or the bodies of the passengers, for the upper apartment 103 feet, and for the lower 112 feet. This vessel on her last homeward voyage lost 100 passengers at sea.

In my opinion not less than 250 or 300 cubic feet should be given to each passenger.

The condition of the passengers at the time of embarkation is a matter of much importance.

By section 41 of the British law now in force, a medical inspection is required of the passengers, either on shore or on board, before sailing. There is strong reason to believe that this inspection is so hastily and superficially conducted, as practically to amount to no protection against the admission into the ship of persons already infected with objectionable diseases. I speak on this point with confidence, having been an eye-witness of the mode in which the examination is sometimes, at least, performed in Liverpool. The inspection I saw was conducted through the window of a little office, and consisted in nothing but looking at the tongues of the passengers, as they presented themselves in rapid succession before the examiner, and exhibited their passage tickets. There was no certainty even that the person presenting the ticket was the passenger named on it, and that its real owner was not, at the moment, laboring under small pox, or typhus fever, in some other place, and was represented at the examining office by a conniving friend.

The examination is said now to be conducted chiefly on board ship, just before sailing, when, amid the excitement and confusion of a crowd, with their luggage and preparations for sea, it may well be conceived that a person who knows he will be ordered on shore, should his disease be detected, will employ every artifice to avoid it.

It is manifest that we, on this side of the Atlantic, are more directly interested in the healthful condition of the hitherward passengers, and therefore this inspection should be our duty rather than of those whose shores the emigrant is about to leave. Our institutions, with their officers and employés, and our population, present and prospective, are the sufferers by the introduction of diseases into the vessels, and upon our coasts, and not those which have been left behind.

Besides the inspection of passengers on account of their bodily diseases, there should be one also with reference to their clothing, and other personal necessaries for the voyage. In numerous instances these people have not only no bedding of any kind, being obliged to sleep on bare boards, but in greater numbers still, they have no change of clothing, and what they have on is already reeking with long collected filth. They step ashore, if so happy as to reach it, in the same unchanged vesture in which they embarked. Such total unpreparedness for an Atlantic voyage must necessarily be productive of suffering and disaster. In cold or boisterous weather, and at other times, in order to avoid observation, they are compelled to keep below, much more than they would had they more decent outfits. The propriety of requiring a certain amount of preparation for the voyage by each passenger, is therefore a point of considerable importance in connexion with the prevention of disease; and the duty of inspection, for all purposes, should be performed by a well qualified surgeon attached to the ship, who should be responsible to American as well as British law. It is on this side of the Atlantic only that the neglect of an officer's duty, or the violation of a passenger's sanitary rights, can be properly made known, and evidence of it produced.

The authority given the captain by section fifth of the act of 1848, to maintain "good discipline and such habits of cleanliness among the passengers as will tend to the preservation and promotion of health," is of the highest importance. It is well understood that the greatest obstructions to the enforcement of this authority are presented by the passengers themselves, and it is said that when attempts have been made to execute the needful measures, difficulties have arisen from their wilful obstinacy and opposition, and that vexatious suits for assault and battery have been instituted against the masters on arrival. That the authority of the captain should be paramount in the enforcement of regulations for this purpose is indisputable, and he should be protected against the revengeful disposition of passengers by the admission, in evidence, as exculpation, of the order or advice of the surgeon given at the time, of the necessity of the enforcement of sanitary measures.

The foregoing statements and suggestions were prepared prior to the receipt of your circular of December 29, 1853; and in looking over the subjects upon which information and opinion are sought, I find that most of them, especially the first six, and the tenth and thirteenth, have been discussed in the preceding pages, although not in the order in which they are laid down in the circular. On the other propositions of the circular, I will therefore add a few words.

With regard to the employment of qualified and experienced surgeons on emigrant ships, I can speak from considerable observation obtained by three years of intimate relationship with emigrants arriving at this port, and I have no hesitation in declaring it, what indeed seems self-evident, a matter of the highest importance. I have been cognizant of frequent instances in which the services of a well-qualified physician were greatly needed and could not be had, and I can readily understand that they would be continually in demand for the care of the steerage, in gross and in detail, in preventing as well as relieving sickness.

But they should be men of much better stamp, of both moral and professional character, than a great many whom I have seen employed for this responsible duty. Surgeons on emigrant vessels are too often appointed on the same principle as in a militia corps—merely to fill up the staff, for which purpose a mechanic or a cook's mate will answer. Certainly an equal degree of qualification should be required in the surgeon as in the pilot; the responsibility of the former, as far as the safety of the passengers is concerned, being as great as that of the latter, whose vocation is only open to those qualified by a long apprenticeship, and proved by examination.

A similar train of thought arises in considering the eighth proposition of the circular.

The ninth proposition is rather a question of morals than of medicine, but it must forcibly strike every one as of great propriety. While, I believe, our own laws are silent on the subject of separate accommodations for the sexes, the English law enforces it; but it is evident that, to be of real value, the arrangements for such separation must be maintained when out of reach of British law, which is, in this case, when out of sight of British soil. It is *here*, if anywhere, that complaints of violation of such a law will be heard, and here should opportunities for redress and punishment, therefore, be afforded.

Of the eleventh and twelfth propositions I may express my unqualified approval. Their adoption would lead to many very valuable statistical results of sanitary, moral, and political character. The eleventh proposition, particularly, I regard as valuable, and, indeed, necessary, should it be determined to require a disinfection of every vessel upon which ship fever or other infectious disease may have appeared.

With regard to the last proposition, I have been less familiar, by experience, with the points upon whose merits the question involved in it should be decided. It is generally considered that the circumstances of tropical latitudes are such as not to justify so much crowding as may be allowed in higher latitudes; but the difference of character between the passengers within the tropics and those between Europe and the United States, which we have been considering, is generally so great as to cause them to be regarded in essentially different aspects in relation to this question.

I have thus, though I fear too prolixly, frankly given you my views on this subject, which, from the magnitude of its interests, both pecuniary and humanitarian, is eminently worthy the attention of our country's legislature, to whom all other nations look for new steps in the improvement of the condition of the race.

For your efforts to make more clean and safe, for health and life, the great highway between them and us, and especially should your efforts be crowned with success, the children of the present and future generations will rise up and call you blessed.

With high regard, very truly and respectfully,

JNO. H. GRISCOM.

Hon. HAMILTON FISH,
Chairman special committee, U. S. Senate.

Extract from the transactions of the New York Academy of Medicine, referred to in the foregoing communication of Dr. Griscom.

LETTER FROM MR. PARKER.

PERTH AMBOY, NEW JERSEY,
March 15, 1852.

DEAR SIR: Having read your treatise on the "Uses and Abuses of Air," I send you an account of what occurred in this place some years since, and which proves the efficacy of fresh and pure air, not only in preventing, but curing disease.

In the month of August, 1837, a number of ships, with emigrant passengers, arrived at Perth Amboy, from Liverpool, and other ports, on board of some of which ship fever prevailed. There was no hospital or other accommodations in the town, in which the sick could be placed, and no person could admit them into private dwellings, fearing the infection of the fever. They could not be left on board the ships. An arrangement was made to land the sick passengers, and place them in an open wood, adjacent to a large spring of water, about a mile and a half from town. Rough shanties, floored with boards, and covered with sails, were erected, and thirty-six patients were taken from on board ship with boats, landed as near the spring as they could get and carried in wagons to the encampment (as it was called) under the influence of a hot sun, in the month of August. Of the thirty-six first named, twelve were insensible, in the last stage of fever, and not expected to live twenty-four hours.

The day after landing there was a heavy rain, and the shanties affording no protection with their "sail" roofs, the sick were found the next morning wet, and their bedding, such as it was, drenched with the rain. It was replaced with such articles as could be collected from the charity of the inhabitants. The number of the encampment was increased by new subjects to the amount of eighty-two in all.

On board the ship, which was cleansed after landing the passengers, *four* of the crew were taken with ship fever, and two of them died. Some of the nurses at the encampment were taken sick, but recovered. Of the whole number of eighty-two passengers removed from the ship, *not one died.* Pure air, good water, and perhaps the rain, (though only the first thirty-six were affected by it,) seem to have effected the cure.

No report has been made of these circumstances, and I send this from my recollection, and the information derived from the physician. Dr. Charles M. Smith, who still resides here, and to whom I refer you.

Very respectfully,
JAMES PARKER.

DR. JOHN H. GRISCOM.

A few further particulars of this case have since been derived from a statement of Dr. C. McKnight Smith, the gentleman referred to by Mr. Parker.

The ship was the Phœbe, with between three and four hundred passengers; a number of them had died on the passage. The shanties spoken of, were two in number, thirty feet long, twenty feet wide, boarded on three sides about four feet up, and over them old sails were stretched. Of the twelve who were removed from the ship in a state of insensibility, such appeared the hopelessness of their condition, that the overseer, who is a carpenter, observed, "Well, Doctor, I think I shall have some boxes to make before many hours." "The night after their arrival at the encampment," says Dr. Smith, "we had a violent thunder gust, accompanied by torrents of rain; on visiting them the following morning, the clothes of all were saturated with water; in other words, they had had a thorough ablution; this, doubtless, was a most fortunate circumstance. The medical treatment was exceedingly simple, consisting, in the main, of an occasional laxative or enema, vegetable acids, and bitters; wine was liberally administered, together with the free use of cold water, buttermilk, and animal broths." The four sailors who sickened after the arrival of the vessel were removed to the room of an ordinary dwelling-house; the medical treatment in their case was precisely similar, yet two of them died. Two of the number suffered from carbuncle while convalescing. The Dr. adds: "My opinion is, that had the eighty-two treated at the encampment been placed in a common hospital, many of them would also have fallen victims. I do not attribute their recovery so much to the remedies administered, as to the circumstances in which they were placed; in other words, a good washing to begin with, and an abundance of fresh air."

No. 5.

Communication from Gregory Dillon, esq., president of the Irish Emigrant Society of New York.

Office Irish Emigrant Society,
51 *Chambers Street, N. Y., January* 14, 1854.

Dear Sir: I have been honored with the receipt of your communication of the 29th ult., in relation to emigrant passengers, which should have received an instant reply, were it not for a short temporary absence from the city.

The subject is a most important one, requiring minute and patient investigation, and I congratulate the emigrant, and the friends of the emigrant, that the subject has fallen into such competent and devoted hands, as I feel persuaded that every justice will be done it from the queries embraced in your circular, showing your familiarity with all its features.

1. There is one point which has struck me as requiring prompt remedy. I refer to double-decked passenger ships, deprived, as they must be, of proper ventilation, the lower deck consequently infecting the upper deck with its impure effluvia.

2. To secure a *sufficient space* for *each passenger* without reference to the tonnage of the ship or the number of persons on board, although the whole number should be limited to the extent of the tonnage.

3. A separation of the sexes, and strict prevention of unnecessary intercourse between the crew and passengers.

Permit me to take the liberty of enclosing to you, herewith, a printed copy of a letter from an intelligent passenger on board an emigrant ship, and first published in the Dublin Nation, of December 20, 1853, giving a graphic description of the sufferings and exposure of those poor people during the voyage.

I will not annoy you with any further allusions to this subject. But beg leave to present my own thanks, as I am also authorized to do so in the name of our society, for your generous zeal and sympathy for our poor country people.

I am, dear sir, with the highest respect, your most humble servant,

GREGORY DILLON, *President.*

Hon. HAMILTON FISH,
United States Senator, Washington city, D. C.

No. 6.

Communication from the German Society of New York.

The great mortality on board of emigrant vessels has occupied for sometime the attention of the directors of the German Society of this city, in consequence whereof they appointed a committee, to investigate, as far as possible, this calamity, in order to propose some remedies, which were to be presented to your honor, in a memorial, as soon as prepared. While occupied with this investigation, this committee met with you circular, dated Washington, 29th December, 1853, and the German Society permits itself to answer the various points as follows: To point—

I. Instead of superficial feet, we hold the opinion that a measuremen t of cubic feet should be adopted, and we place here a space-table, giving to each passenger as follows:

1. In ships which have but one steerage: if the ship has from 1 to 100 passengers, each passenger to have 100 cubic feet.

From 101 to 120 passengers, each passenger to have			- -	101
121 to 140	do.	do.	do. - - -	102
141 to 160	do.	do.	do. - - -	103
161 to 180	do.	do.	do. - - -	104
181 to 200	do.	do.	do. - - -	105

cubic feet, and, in the above manner, adding for each additional twenty passengers one cubic foot to every passenger on board.

2. In ships which have two steerages—that is, upper and lower:

(*a*.) If the lower steerage has from

1 to 100 passengers, each passenger to have			- -	150	cubic feet.
101 to 120	do.	do.	do. - -	152	do.
121 to 140	do.	do.	do. - -	154	do.
141 to 160	do.	do.	do. - -	156	do.
161 to 180	do.	do.	do. - -	158	do.
181 to 200	do.	do.	do. - -	160	do.

And, in the above manner, giving to each passenger on board additional two cubic feet for every twenty additional passengers.

(*b.*) If the upper steerage has from—

1 to 100	passengers, each passenger to have			- -	110 cubic feet.
101 to 120	do.	do.	do.	- -	111½ do.
121 to 140	do.	do.	do.	- -	113 do.
141 to 160	do.	do.	do.	- -	114½ do.
161 to 180	do.	do.	do.	- -	116 do.
181 to 200	do.	do.	do.	- -	117½ do.

And, in the above manner, giving each passenger on board an additional one-and-a-half cubic feet for every twenty additional passengers.

In no instance should a ship with a single steerage carry more than five hundred passengers, and ships with two steerages should never be allowed to carry more than two hundred passengers in the lower steerage and more than four hundred in the upper steerage. This space we presume to be sufficient to make the berths six feet long and twenty-one inches wide for each passenger, instead of eighteen inches wide, as provided in chapter xvi, section 3, passed 22d February, 1847.

II. Chapter xli, section 4, passed May 17, 1848, does, to all appearance, not provide sufficiently for this point; the following might be an improvement: 25 lbs. of navy bread, 10 lbs. of rice, 5 lbs. of oatmeal, 5 lbs. of wheat flour, 10 lbs. of peas and beans, 40 lbs. of potatoes, 1 pint of vinegar, 60 gallons of fresh water, 15 lbs. of salt pork, free of bone; ½ lb. of black tea, 2 lbs. of butter, 1 gallon of molasses, salt at discretion, all to be of good quality, and a sufficient supply of fuel for cooking; but at places, etc., etc., as the paragraph proceeds now.

III. To this point we would reply, we have found by inquiries during the last years, that passengers are best served and taken care of where the captain provides the food, ready prepared and cooked, for the passengers; however, in case the captain furnishes the food ready cooked, there should always be on board one camboose or cooking range, the dimensions of which to be six feet long and two feet wide, for the special use of such steerage passengers who desire to cook such little articles of provisions as they have brought with themselves for their comfort. We have often heard bitter complaints, in cases where the passengers were obliged to cook their own victuals, that on account of the insufficient cooking conveniences for the great number of passengers on board, they were deprived for days in succession of warm or freshly cooked food; and it really deserves particular inquiry and recommendation, that, in every instance, the ship should provide the meals ready prepared, and the master should be placed under proper regulations how and what to give for each meal.

IV. Chapter xli, section 2, provides for this; it might be added, yet, however, that in case a ship carries more than three hundred passengers, the steerage should have at least three ventilators or hatchways.

V. Chapter xli, section 3, does not provide sufficiently for this point; it might be amended so as to state distinctly, that, in case the passengers are to prepare their victuals themselves, every two hundred passengers shall have one camboose or cooking range, the dimensions of which shall be six feet long and two feet wide; however, we recommend again strongly our answer to point III.

VI. Chapter xii, section 5, provides for this sufficiently, and it is only necessary to have this law properly enforced; we would recommend yet, that the cleaning of the steerage by disinfecting agents should be particularly adopted, and the use of water in scrubbing the steerage entirely abolished, as this causes much damp and unhealthy atmosphere, producing many colds and other sicknesses.

VII. Touches a want which has been felt most keenly on board all emigrant vessels, and in no instance should a ship be allowed to carry more than one hundred passengers, without having an experienced graduated surgeon on board.

VIII. Each one hundred passengers should have one female nurse to attend the sick and see that proper cleanliness is observed by the steerage passengers, for much complaint has heretofore existed against the brutal language and behaviour of the crew towards the sick.

IX. The separation of the sexes, in so far as it concerns unmarried adults, is one of the most necessary points your honorable body should provide for, especially as it is one of such easy accomplishment; and in no instance should any of the crew have access to the female department, unless it be in company with the captain or first mate; and in case there are more than one female nurse on board, then at least one nurse should sleep in, and take charge of, this apartment. It might be well to state here, that it shall be voluntary for single females to enter this so arranged apartment, or remain with their friends or families in the general steerage; so that, by this clause, families or friends could not be separated against their will.

X. This is already partly answered in point VI.

XI and XII. The captain of each vessel bringing emigrants should be obliged to make out a separate and correct manifest of such passengers who have died on board his ship during the voyage, stating the full name, age, last place of residence, and day of death, also the kind of disease; this should be accompanied by a certificate for each death, signed by captain, first mate, surgeon, and at least five of the male passengers, being twenty-one years of age.

XIII. This point is already answered by point I.

XIV. It is our opinion that vessels passing the tropics should give passengers fifty per cent. more room than those given as stated in point I.

We permit ourselves to suggest further:

Chapter XII, section 7, provides that an inspector from the customs shall visit each passenger vessel and report in writing to the collector, if the laws regarding emigrant passengers have been complied with; this law should be more strictly enforced, and we would most humbly suggest, that in all the principal ports of the United States, where passengers arrive, a separate special inspector be appointed, who speaks both the English and German languages, whose office it is to visit, at arrival, and before discharge of passengers, each emigrant passenger vessel, and there inquire of the passengers if, and how far, the various laws of the United States have been complied with, and, in case complaints are made, to accept the same from the passengers under oath, on board, upon which he is to make his immediate report to the collector of the customs and the commissioners of emigration. This would, no doubt, procure for the passengers better accommodations—health on arrival in a foreign land, and render them more fit to look at once for proper employment; while under the present system they have too often filled our hospitals and formed pitiful, wretched, and haggard looking beggars in the streets; and it must not be overlooked, that not only the mortality consists of those who die on ship-board; no! we must also regard those thousands who arrive here emaciated, sick, and helpless, and shortly after die in our hospitals.

By a proper observance of such amended laws, we would properly cause an advance in the price of passage; this, however, would make no loss to either the passenger or this country; for the first would gain more in the preservation of his health than the value of a few dollars, and the last would, in all probability, get just as many of the valuable industrious class of poor emigrants, and no doubt avoid the arrival of paupers, who never could earn anything for themselves in their own country, and help here only to fill our poorhouses and abodes of idleness and drunkenness.

We would also suggest that a small apartment should be arranged for a hospital on board all ships which carry more than one hundred passengers, where such sick passengers could be administered to who need quietude and the particular attention of the surgeon. Sick members of families who prefer to remain with their families should not be obliged to enter the hospital. All the separations in the steerage could be made by slats of wood, so as not to hinder the free circulation of air through the steerage.

R. A. WITTHAUS, *Chairman*,
AUG. MOLLMOMY,
HERM. ROSESS. } *Committee.*

Approved:

JOHN C. ZIMMERMANN, Sen.,
President of the German Society.

Hon. HAMILTON FISH,
Chairman of the Select Committee, &c.

NEW YORK, *January* 11, 1854.

No. 7.

COMMUNICATION FROM LEOPOLD BIERWITH, ESQ., IN ANSWER TO THE CIRCULAR OF THE COMMITTEE.

Answers to queries of the Emigrant Committee of the Senate of the United States.

The space allotted to each passenger by existing laws appears to me large enough in large vessels of one thousand tons and upwards, assuming that in them the between decks are much higher than in smaller vessels. The fourteen or sixteen superficial feet in a vessel of one thousand five hundred tons may contain fifty per cent. more cubic feet than in a vessel of one hundred and fifty tons. It might be well to state the place allotted to each passenger in cubic feet, and not in superficial feet. I do not see the necessity of such a distinction with respect to the number of passengers as suggested in query XIV; but I think vessels should not be allowed to carry as many passengers—in proportion to their capacity—in the winter months as in the summer months; in other words, from November to March, inclusive, the room allotted to each passenger should be larger than from April to October.

The quantity of provisions required by existing laws appears to me sufficient, nor would I find fault with the quality, provided the law were complied with strictly and in good faith; but that will not be the case, and cannot be expected, so long as passengers are permitted to furnish their own provisions, and have to cook them themselves. I am clearly of opinion that suffering on the passage from deficient and indigestible food, resulting in disease and death, will not cease until passenger vessels are obliged by law to provide food for all the people on board, and to cook said food and serve it out at stated periods of the day.

The existing laws respecting ventilation are deemed sufficient, if fairly complied with. In connexion herewith, it might be enacted that passenger vessels shall carry nothing on deck but what may be absolutely necessary for their safety, so as to leave the largest possible room to the passengers for exercise.

Personal cleanliness of the passengers and crew, and cleanliness of the ship, cannot be too rigidly enforced. I would have every passenger go through a bath and clean linen put on before embarkation.

A *qualified* and *experienced* surgeon would be a great comfort, but too costly a luxury on board of an emigrant ship; and, satisfied that most of the surgeons generally employed, and who may be procured for a moderate compensation, will do more harm than good, I would rather have none. Most ship-captains are competent to treat common cases of sickness at sea; and it is a rare thing that among so great a number of passengers as are collected on board of every emigrant vessel, there should not be one acquainted with the healing art, whose services might be made available in urgent cases. Much more essential than a surgeon appears to me the employment of a number of attendants to minister to the sick, to distribute the food at meal times, and to enforce the observance of cleanliness. And I would urge the enactment of a law by which every emigrant vessel should have a steerage steward and a steerage stewardess for every twenty-five passengers on board.

The demoralizing tendency of the existing system is self-evident. A change effecting the separation of the sexes would be highly desirable; Its practicability is proved by the British convict ships.

The report of the particulars of the voyage might be interesting; but its practical use may be questioned, unless the report be made to some public officer, to whom the passenger may resort for redress of grievances. I am satisfied the condition of steerage passengers at sea would be greatly improved, if, at the place of landing, they were to find a public authority to whom they could complain of violations of the passenger laws, with the certainty of obtaining immediate satisfaction. As it is, the passengers may be, and often are, shamefully abused with impunity, because they have no means of obtaining redress without tedious and expensive proceedings. The ill-used emigrant applies to the consul of his nation, or to some society established for his protection; but it is utterly out of their power to follow up every abuse and outrage, nor has the poor plaintiff the means to await the slow decision of court. What is wanted is a tribunal, with power to decide, without lawyers and jury, and to enforce immediate compliance with its decrees. If such a tribunal existed, solely for the purpose of trying cases arising under the passenger laws—if the emigrant could find it at once on landing, and at all times ready to listen to his complaints, and to give immediate effect to its decisions, it would, in great measure, render further legislation respecting the carriage of steerage passengers superfluous—the existing laws would almost prove sufficient.

The present state of things is, in a modified form, a repetition of what was witnessed in 1847–'48, when (as mentioned in a report of the Commissioners of Emigration of the State of New York) the number of persons who perished by ship fever at sea, and in the various emigrant hospitals in American ports, was estimated to exceed 20,000. There is also a remarkable analogy between that period and the present in the high price of bread-stuffs and all other articles of human food; and this confirms me in the opinion expressed years ago, and adhered to ever since, that the mortality at sea is mainly, if not entirely, owing to the want or insufficiency of wholesome nourishment during the passage. The subject at that time attracted the attention of the New York Chamber of Commerce, and a committee of that body, to whom the matter was referred, (a copy of whose report is annexed,) arrived at the conclusion that the main causes of the many deaths at sea were "want of food and want of pure air in the between-decks."—(Extract from the report referred to above.)

Subsequent legislation has been, to a considerable degree, effectual in remedying the latter, but the want of food on the passage still continues; and, until it be made the duty of the ship to provide wholesome and sufficient food for the passengers, to cook said food, and serve it out at stated periods of the day, so long as passengers are allowed to furnish their own provisions, and are obliged to cook them as best they may, there will always be more or less suffering and sickness and deaths on board of emigrant vessels from Europe.

Respectfully, &c.,

LEOPOLD BIERWIRTH.

NEW YORK, *January* 12, 1854.

No. 7.

Communication from Mr. Bierwirth, of New York.

Next to the miserable state of health in which so many emigrants embark, the great causes of the deplorable condition in which they arrive in our ports are—

Want of sufficient and wholesome nourishment during the passage, and want of pure air in the between decks or steerage, where emigrant passengers are generally located.

The Chamber of Commerce is doubtless aware that emigrants from Great Britain and Ireland can claim from the ship that carries them nothing in the shape of food, except one pound of bread and three quarters of a gallon of water each per day; and it is a well-known fact, that even this they cannot always obtain—that the law is shamefully violated—that vessels leave Liverpool without providing the required quantity of bread. But all nourishment, beyond bread and water, must be furnished by the emigrants themselves, and it is their business to get it cooked as best they may. The consequences of this arrangement to the poor, careless, improvident people are self-evident; many of them embark without any provisions of their own, and very few, if any, with a sufficient supply; many have not the means to buy food, and those who have, deceive themselves as to the duration of the voyage; and hence it is doubtless true that not one of all the emigrant ships from British and Irish ports has a sufficient supply of proper food for all on board. But supposing there are some among the cargo of passengers well-provisioned for the voyage, what are the means at their disposal for having their food properly cooked? It has already been stated that they must do their own cooking. What are the arrangements of which they may avail themselves for the purpose? Your committee will describe them, as found on board of the very largest and best of ships engaged in the conveyance of emigrant passengers, leaving the Chamber to imagine what they must be on board of the fleet of vessels of an inferior class.

On the upper deck of the ship there are two small cook-rooms about five feet deep and four feet wide, called the steerage galley. Within is a grate corresponding with the width of the room, over which grate is fastened an iron bar, and on this there are two iron hooks, to which the emigrant hangs his pot or kettle (if he have one) when he wants to cook. These are all the arrangements for preparing meals for several hundreds of passengers! The result is, that, except when they have nothing to cook, or are sick. (which is too often, unfortunately, the case,) there is constant fighting for room near the caboose, and not one of the passengers can be sure of getting his food well cooked.

Thus then the consequences of the existing arrangements are not only an absolute want of sufficient and wholesome food, but also the impossibility of properly preparing what little there may be; and, with this fact before us, it cannot surprise us to see the emigrant, greatly enfeebled already when going on board, either die on the passage, or arrive here with scarcely a spark of life in him. Your committee therefore, with a view of removing one of the great causes of the evil complained of, would urge the enactment of a law requiring vessels bringing emigrant passengers from Europe to our ports, not only to provide food for them, but also to cook said food, and serve it out for them at stated periods of the day.

The objection to such a law urged by a portion of our shipping interest is this: that it would, or might subject the vessel, or captain and owners thereof, to vexatious law-suits on the part of the passengers, complaining as well of quantity as quality of food thus furnished; but this complaint can be anticipated by a clear and explicit statement before commencing the voyage, setting forth the nature and quantity of food the ship agrees to furnish, and the hours of serving the meals each day. It is the object and clear intention of the proposed law that a sufficient quantity of wholesome food, properly cooked, shall be served out every day to every passenger on board, and if this be complied with, there can be no cause for reasonable complaints. And your committee would further remark, that such complaints are not heard of on the part of passengers in vessels coming from Bremen, Hamburg, and other continental ports, where the system recommended by your committee has been in operation for a number of years, to the satisfaction of all concerned. The Bremen regulations as to the quantity of food required on board passenger vessels—a copy of which can easily be procured—might serve as a guide in framing the law.

It is also objected that said law would diminish emigration, because it would increase the cost of the voyage. This objection is deemed futile. No vessel should leave a European port without a sufficient supply of food for all on board, and if this be the rule, it cannot make a difference in the cost of the necessary provisions, whether the vessel provide for all, or each passenger do so for himself. If a difference exists, it must be clearly in favor of the ship, inasmuch as she can buy the articles required in large quantities from the wholesale dealers, and thereby procure them cheaper than the emigrant, who has to go for his small wants to the retail store.

Your committee now come to the last great cause of suffering among emigrant passengers, namely, the want of pure air in the between-decks, or steerage.

The nature of the atmosphere in the steerage of an emigrant ship can readily be imagined, without a minute description. We have only to consider that the room is rarely more than six feet high, has no other aperture for the admission of fresh air than the hatches, which, during night and bad weather, are generally closed, is crowded with passengers, of whom the greater portion are strangers to the virtue of cleanliness, and many of them down with sea-sickness or other equally loathsome diseases. What, with the miasma of a damp hold, the excretions and exhalations from the bodies of the individuals thus confined, and the emanations from other and more offensive matter, an atmosphere is created which must act like poison on those who have to breathe it.

To remedy this evil, the preamble and resolutions offered to the Chamber, recommend the enactment of a law "requiring all vessels engaged in the conveyance of emigrant passengers, to have the necessary arrangements for cooking under deck, with a view of thereby producing ventilation and disinfection;" and also requiring such vessels to introduce "air pipes," for the purpose of supplying the vacuum that necessarily will be occasioned by the action of fire employed in the process of cooking.

It is readily admitted that fire is one of the most effective agents that can be employed in controlling ventilation; but its introduction below deck, into the midst of the passengers, suggests so many serious objections, that your committee cannot agree to recommend the plan to your favorable consideration. Apart from smoke and steam, which it would be very difficult, if not impossible, to prevent from spreading below deck, to the great inconvenience of passengers, the danger attending the plan is deemed so great, that shipmasters positively refuse encountering it, and the board of underwriters, to whom the question was submitted, whether it would affect the marine risks of vessels having the arrangement for cooking below deck, reply, that after a full discussion of the subject, it was unanimously resolved:

"That, in the opinion of this board, the hazard would be greatly enhanced."

But your committee are willing to believe it does not require the assistance of a dangerous agent to insure at all times during the voyage a full supply of fresh air in the between-decks, or steerage; that it can be obtained by means of iron tubes and other contrivances. This belief does not rest on theory alone, but also on experience. The plan which your committee wish to recommend, was introduced several years ago by Capt. Jos. C. Delano, with highly satisfactory results. It has since been improved by Capt. E. Knight, and is now in operation on board of several of the Liverpool and London packet-ships. It is extremely simple.

Four or six iron tubes, projecting above deck, so as to be under the immediate action of the wind, are led down into the between-decks. One-half or one-third of the number are for receiving fresh air, the others for discharging the impure air; the former have receiving caps, the latter exhausting caps. These tubes can be applied to any vessel, small or large, and they always act in proportion to the strength of the wind, and hence are most efficient when most needed—say in stormy weather, when hatches are closed.

In addition to these air pipes, all vessels engaged in the conveyance of emigrants should be required to have substantial houses built over the hatches used by the passengers. These houses should have a door each side, to open about twelve inches from the deck, that one may be kept open all weather.

With these arrangements, a sufficient supply of fresh air in the between-decks can no doubt be insured, and your committee would earnestly recommend the adoption of a law requiring the same to be generally introduced.

It is proper to add, that to such a law there would be no objection on the part of our shipping interest; on the contrary, your committee have the satisfaction to state that our principal ship-owners are desirous to have the law passed, and have already communicated their wishes and views to Washington.

Independently of the measures suggested, others might be adopted with salutary effect, but your committee deem it inexpedient to embrace them in the proposals for legislative action, because it would be difficult to enforce laws concerning the same.

Thus it would no doubt be desirable to have an officer on board of every emigrant ship, whose sole business it should be to enforce cleanliness on the part of the passengers, and to see to it that no offensive matter be suffered to remain in the between-decks, or steerage. Disinfecting agents should be freely used; and, among these, Le Doyen's disinfecting liquid may be recommended. It has been tried with great success in the institutions under the control of the commissioners of emigration, and the superintending physician of one of them writes: "It certainly possesses the power of destroying offensive odors effectually, and without leaving any smell of its own." As auxiliaries to the air pipe, canvas windsails might be used in moderate weather; and it is also suggested that a simple fan-wheel or blower, placed between decks, with a flexible canvas cylinder leading from it to deck, if properly used, would aid greatly in changing the poisonous atmosphere of the steerage.

In conclusion, your committee would urge on the Chamber of Commerce the expediency of asking the government of the United States, at Washington, promptly to call the attention of the governments in Europe, more particularly the one of Great Britain and Ireland, to the important subject under consideration, so that the proposed laws, if enacted, may be met by corresponding regulations in the ports of Europe.

CHAS. AUG. DAVIS, Chairman, } *Committee.*
LEOPOLD BIERWIRTH, }

NEW YORK, *March* 7, 1848.

No. 8.

Letter from R. B. Minturn, in answer to the circular of the committee, and enclosing letters from Captains Knight and Britton, experienced commanders of passenger ships.

NEW YORK, *January* 9, 1854.

MY DEAR SIR: I have thought that I could best answer the inquiries in your circular by sending you the opinions of two of our most experienced captains, which you will find enclosed, addressed to me. They are both men of excellent judgment, and thoroughly informed on the subject.

They think that *cleanliness* is the greatest preservation of health to the emigrants, but it is very difficult to enforce it; and I think their suggestion of aiding them in this respect by the authorities is very judicious.

Yours, very truly,

R. B. MINTURN.

Hon. HAMILTON FISH.

NEW YORK, *January* 6, 1854.

SIR: In answer to the circular from the Hon. Hamilton Fish, which you handed us yesterday, we will take up the queries in the order in which they are proposed, and very briefly give you our views. And here allow us to say that we are obliged to conform in every respect to the late act of the British Parliament for regulating the carrying of passengers both to and from Great Britain, and any act passed by our government conflicting in any respect with the British act will be of no service to us, and may do us great injury.

1st. The space allotted to each passenger. Our laws allow us to carry an adult to every fourteen superficial feet measured on deck, while the British act allows one adult to be carried to every twelve superficial feet. We can carry passengers according to our laws very comfortably.

2d. The quantity and quality of provisions, &c. These are provided by the English act, and are sufficient in quantity, and pass a rigid inspection from the emigration agents there before being put on board.

3d. In answer to the third query, we would say that the master should in all cases provide for the passengers.

4th. Ventilation. The more the better. The ships now in the European trade are very well ventilated.

5th. Cooking arrangements. Are sufficient at present as ordered by the English act.

6th. Cleanliness, &c. This is all important. At present we have no legal power whatever to enforce sanitary regulations. Power should be given to commanders of ships to enforce obedience to necessary regulations for daily cleanliness. We think this could be the most easily accomplished by giving them power to stop the supplies of provisions and water to such persons who refused to comply with such regulations.

7th. Surgeons. These are required by the English act for every ship carrying over five hundred passengers, and, in our opinion, is as it should be.

8th. Employment of attendants for the passengers, &c. One passenger steward and two passenger cooks are required by the English act, and are sufficient, with the surgeon and ordinary ship's officers, to look after the passengers.

9th. Separation of sexes, &c. This is also provided for by the English act, and is carried out as far as possible without separating members of the same family. The crew are never allowed between decks at all, except on ship's duty.

10th. Disinfecting. This we always do at sea every day, as far as possible, using for the purpose chloride of lime, disinfecting fluid, and the smoke of tar and sulphur.

11th. Report of vessels on arrival, &c. This is always given to the health officer, who boards us at quarantine on arrival.

12th. Inquest by federal officers. Impracticable.

13th. Limitation of the number of the passengers in proportion to the registered tonnage of vessels. This would be unjust and injurious to emigrants, as the registered tonnage of a vessel is no index of her *real capacity.* The actual space on board is the only correct rule for determining the number of passengers a ship is entitled to carry. By the rule of tonnage, a clipper ship would be allowed double the number that she would be entitled to under the present regulation, or would be able to accommodate properly.

14th. Distinction in respect to vessels passing between the tropics. This would be proper, and is also provided for in the English act. More room should be given in warm climates than in cold.

Yours, respectfully,

HALE S. KNIGHT, *Ship New World.*
JOHN BRITTON, *Ship Constitution.*

ROBERT B. MINTURN, Esq.

No. 9.

Letter from Cyrus Curtis, esq., one of the Commissioners of Emigration of New York.

NEW YORK, *January* 20, 1854.

SIR: I have the honor to acknowledge your favor of the 29th ultimo, containing a series of questions touching our present emigration laws. In order to correct certain abuses that exist under our own laws, something at least may be learned of England, who is somewhat in advance of us, and whose laws are, if possible, too stringent—so much so as to check emigration in some cases.

I am not aware that I can give you any new information on this subject, but beg herewith to transmit such as I have collected in answer to your interrogatories.

With great respect, I remain your obedient servant,

CYRUS CURTIS.

Hon. HAMILTON FISH,
United States Senate.

I. Quite sufficient. See answer XIII.

II. Inasmuch as epidemic diseases have been almost entirely banished from our navy, and the exemption of sailors and marines from these diseases may, in a great measure, be attributed to the quantity and quality of their food, it might be well to adopt the diet list of our navy, as established by Congress in 1842, or the following table, which is a modification thereof, and which, as it excludes liquors and embraces potatoes, an almost indispensable article of diet with most emigrants, may be considered preferable.

The English law provides that each steerage passenger shall be provided with a contract ticket, stipulating the quantity of provision that each passenger shall be served by the master.

Table—full diet.	Pounds.							Ounces.						Fract'ns of a pint.	
	Beef.	Pork.	Salt fish.	Potatoes.	Rice.	Beans.	Pickles.	Biscuit.	Sugar.	Tea.	Coffee.	Butter.	Cheese.	Molasses.	Vinegar.
Sunday	1	....	....	½	....	½	....	14	1	⅛	½	....	....	....	....
Monday	....	1	....	½	....	½	....	14	1	⅛	½	....	....	....	....
Tuesday	1	....	....	½	....	½	¼	14	1	⅛	½	2	2	....	....
Wednesday	....	1	....	....	½	....	....	14	1	⅛	½	....	....	¼	....
Thursday	1	....	....	½	½	½	....	14	1	⅛	½	....	....	....	....
Friday	....	....	¾	½	....	½	....	14	1	⅛	½	2	2	....	¼
Saturday	1	....	....	....	½	....	¼	14	1	⅛	½	....	....	¼	....
Weekly quantity	4	2	¾	2½	1½	2½	½	98	7	⅞	3½	4	4	½	¼

Oatmeal and Indian meal are omitted, from their tendency, if used habitually, to induce diarrhœa and dysentery. Vinegar and pickles are anti-scorbutic.

III. No passengers should be allowed to provide for themselves; but in all cases it should be the duty of the master of the vessel to furnish the provisions.

IV. A sufficient supply of fresh air is, of all sanitary matters, of the most importance. To a deficiency of this supply is also attributable the vast mortality among emigrants while crossing the Atlantic, and even during shorter transits, such as was exemplified on board of a ship crossing from Ireland to Liverpool, when, on the 2d December, 1848, seventy-three individuals perished. Hence the question, how is a sufficient supply of pure air to be furnished on ship-board? is the most important of the series.

V. The cooking apparatus and arrangements should be sufficiently ample to cook the meals of all the passengers at regular periods. The passengers should not be permitted to cook their own victuals.

VI. Thorough subordination should be maintained by strict discipline, and the power of the master should be absolute. Personal cleanliness, as well as that of the vesel, should be rigidly enforced; and, if practicable, an hospital for the sick should be located apart from the dormitories of the healthy passengers.

VII. The laws of England require the owners of ships to furnish medicine, and every ship having on board 100 persons or upwards to be provided with a surgeon. The like requisition for American vessels is equally important. English surgeons are generally qualified to discharge the duties of physicians.

VIII. Nurses and other attendants on the sick might be selected and appointed from among the most intelligent emigrants by the master of the vessel, who would be willing to render their services for a small pecuniary consideration or perquisites.

IX. Children of both sexes, of ten years old and under, might be permitted to remain with their, or under the care and supervision of the mother; but we have reason to believe that chastity, as well as decency, requires a separation of the males from the females.

X. There is no process of disinfection so thorough as cleanliness and free ventilation. To destroy noxious odours, *Le Doyen's Disenfectant* is to be preferred to others, for the reason that it instantaneously destroys offensive smells without substituting any peculiar odour of its own. Chloride of lime produces an effluvium not only very disagreeable, but pungent, irritating the eyes and air passages, and often producing a distressing cough; so with sulphur. Whitewash purifies for the time being, by absorbing the noxious vapor, or miasma; but the lime retains as well as absorbs them, and will continue to retain them for months and years, and may also evolve them and reproduce disease under a confined and exalted temperature. Boiled linseed oil, when brushed warm into the planks and floors, makes a very good coating, which prevents, to a very great extent, the absorption, by the pores of the wood, of poisonous exhalations. It facilitates also the cleaning of the floors, which, after its application, assume an oaken color, and diminishes the dampness which is apt to remain a long time between decks after washing or scrubbing. It is, however, next to an impossibility to disinfect a foul ship so that disease will not recur in it at some future period. Fumigation is of very little utility, excepting so far as it invites a fresh current of air. Some few years since it was found necessary to let water into the United States ship Ohio, by scuttling her, in order to eradicate the yellow fever. In most cases, cleanliness and ventilation will speedily extinguish an epidemic.

XI. This should never be neglected.

XII. Let that be made the duty of the surgeon, setting forth the cause of death, whether by disease, injury, neglect, or from other cause.

XIII. It would be preferable to limit the number of passengers according to the space they are to occupy, instead of the tonnage of the vessel; and if there be three decks to the ship, the lower one should never be occupied by them. Every adult passenger should be allowed to sleep alone, or, at all events, no more than two should be allowed to occupy the same berth.

XIV. With proper ventilation, the present space allowed is ample anywhere and for any voyage.

No. 10.

Letter from Messrs. Oelrichs & Co.

NEW YORK, *January* 20, 1854.

SIR: In reply to the inquiries as to the adequacy of the existing laws with reference to emigrant vessels, contained in your circular letter of December 29, we beg leave herewith to lay before you our views on the subject.

1. We consider the space of fourteen superficial feet, now allotted to each passenger, to be sufficient in vessels of one passenger deck, but would recommend an increase of space in vessels of two passenger decks; because, in the first place, the ventilation of the lower decks in vessels of this class must necessarily be imperfect; and, in the second place, these vessels now carry too many passengers in proportion to the space of the *upper* deck, which, in fine weather, would interfere

with their comfort and health, by restricting the room necessary for exercise and the enjoyment of fresh air.

2 and 3. We have no doubt that the enactment of a law making it the duty of the master of the vessel, in all cases, to furnish the provisions and provide for their cooking, would be found highly beneficial, and deem this subject of great importance, inasmuch as it would not only contribute, probably in a greater degree than almost any other measure, to the health and comfort of passengers, but would, besides, by increasing the cost of transporting them, act as a bar to the immigration of those worthless paupers who only fill our almshouses and penitentiaries; whilst, on the other side, it would tend to attract to our shores many of the better class. This system has, many years ago, been adopted by the city of Bremen, and enforced on all emigrant ships leaving its port. From our own experience, we know that it has worked well, and should consider a perusal of the Bremen regulations, which can be furnished by our friend the Hon. R. Schleiden, worthy of the attention of the honorable committee, in case it should see fit to adopt any measures of that kind.

4 and 5. On the subjects of ventilation and cooking arrangements, we have not sufficient data on which to form an opinion as to the adequacy of existing regulations.

6. The master should be responsible for cleanliness of the vessel, but the enforcement of personal cleanliness, however desirable, would, we fear, meet with too many obstacles on the part of the passengers themselves, and give rise to many complaints of tyrannical and brutal treatment.

7. The difficulty in determining the competency of a surgeon, and in obtaining one of requisite experience, would be a serious objection to a law requiring their employment. It would, however, be well to oblige every passenger ship to be provided with a suitable medicine chest, containing such simple remedies for contagious diseases as could be applied with safety by an inexperienced person.

8. In cases of sickness the master could employ some of the passengers as attendants.

9. Separation of the sexes would be advisable, but such separation should not be enforced on members of one family. The crew should be entirely excluded from the apartment allotted to the females.

10. Thorough disinfecting of every vessel on board of which a contagious disease has made its appearance, would be advisable.

11 and 12. The provisions of these paragraphs we also consider quite desirable.

13 and 14. We would suggest a limitation of the number of passengers allowed to every vessel, to two for every five tons register, and an increase of space when a vessel passes within the tropics.

The recent mortality on board of emigrant vessels was, we believe, mainly the natural development of disease contracted before embarkation, and not owing to want of attention on the part of the master or his subordinates.

We remain, with respect, sir, your obedient servants,

OELRICHS & CO.

Hon. HAMILTON FISH,

Senate Chamber, Washington.

No. 11.

Paper received from Mr. Rucker, late Minister from Hamburg to Prussia.

The recent alarming mortality on board emigrant vessels from Europe calls for the attention of the legislature. It has been stated that during the month of November last, 28 of the whole number of emigrant ships which arrived at New York had cholera on board, and that of 13,762 passengers, no less than 1,141 died by this disease, and between four and five thousand were afflicted with it during the passage.

This dreadful amount of suffering and death may have been the consequence of some special epidemic, extending over certain latitudes of the Atlantic, breaking out among the passengers when the ship enters these regions. Yet, if such an epidemic existed on the ocean, it would equally have infected other passenger vessels and steamboats running on the same course. This, however, does not appear to have been the case, as only emigrant ships offer such awful scenes of mortality, and even among them several have arrived during the same period from Europe in an entirely healthy condition. Besides, although the mortality has been in the present year more than ordinarily fatal, the reports of the commissioners of emigration at New York show the number of diseased immigrants to be regularly excessively large at certain times of the year; their hospitals are crowded annually during the winter season with numbers of newly arrived emigrants affected by typhus fever, and the severest forms of disease contracted on ship-board. Great distress seems, more or less, to mark every approach of the winter season. The commissioners attribute this bad state of health chiefly to the feeble condition of a large number of the passengers, arising from the privation they suffer during the sea voyage. It has also been stated that some emigrant vessels are known to be constantly infected with typhus fever, so that this frightful disease breaks out regularly on every successive voyage among their unhappy passengers. One of the physicians of the marine hospital assured us that, whenever the arrival of certain such vessels was announced to him, he could always know beforehand, with certainty, that it would bring him a large number of cases of that kind.

Such a state of things requires a close investigation, and impresses the duty of providing, as far as possible, for this helpless class of persons who, seeking a home in this country, are entitled to the protection of her laws. We are glad to see that the Senate of the United States has already, on the motion of the Hon. Hamilton Fish, appointed a select committee to consider the causes of the sickness and mortality prevailing on board the emigrant ships, and whether any, and what, further legislation is needed for the better protection of the health and lives of passengers on board such vessels.

It may be of interest to compare the present state of the legislation of the United States, in this respect, to the provisions made by other States. We select for this purpose the legislation of Germany, from whence, during the last year, the greatest amount of passengers have come to this country. Among the German States, Bremen and Hamburg are the principal ports for the shipping of emigrants; the laws of these two States relating to the carriage of passengers have been considered as providing best for the protection of the emigrants, and have, for this reason, been taken as a model by the rest of Germany. The principal objects of these laws are the following, viz:

1. To admit to this sort of business only such persons who, by their character, offer a certain guaranty for their respectability; at Bremen, only those who have the full merchants' rights are allowed to carry passengers in merchants' vessels. Besides, they must deposite with the board of control the sum of $5,000, as a security for the strict compliance with the provisions of the laws, and they must have been approved by the board of control.

2. The regulations as to the space required for each passenger on the vessels are, for those ships that are destined to go to the United States, the same as have been enacted there; for other countries, there shall be allowed to each passenger 12 superficial feet of the passengers' deck, while the American act of May 17, 1848, requires 14 feet on the lower deck, and 30 feet on the orlop-deck; if the height of the decks is less than 6 feet, it requires 16 feet; if it is less than 5 feet, it requires 22 feet. The American laws allot, as we see hereby, even a larger space than the German laws, which deem it more essential to provide, in other respects, efficaciously for the comfort and material welfare of the passengers, as we intend to show hereafter, than to give them a still larger space. It is impossible to accommodate emigrants equally spaciously, as may be done with cabin passengers, who pay three times the amount of passage-money; and we should think that, with proper regard for ventilation and purification of the air by fumigating, as the German laws prescribe, the present allowance of space is quite sufficient.

The German laws are much more explicit and careful in regard to the general condition of the vessels, which, perhaps, is of much greater importance. It is epecially enacted that the passengers deck must be unoccupied by goods, and that no goods shall be stowed between the berths of the passengers. The berths are to be dry and comfortable, 6 feet long by 18 inches broad; there shall be no more than two tiers, and the under berths shall be at least 4 inches above the deck; the height between the decks must be at least 6 feet, and the deck covering shall be at least one inch and a half thick. Several other points are also provided by the German laws, which contribute materially to the comfort and health of the passengers, relating to the ventilation and cleanliness, to the cooking, to the amount of fuel; and it is, for instance, especially enacted that two lanterns between the passengers' deck are to be kept burning from dusk till day-light.

3. The principal care of the German legislation refers to a proper supply of food, and it will be seen from the accompanying table, comparing the amount required by the laws of the different countries, how much better, in this respect, the laws provide for the passenger at Bremen and Hamburg than in other ports.

The following supply of food is required for each passenger on vessels going to the ports of the United States from the below-mentioned ports, (the different weights being reduced to pounds of Bremen weight:)

Articles.	United States of America.	Bremen.	Hamburg.	Havre.	Antwerp.	Rotterdam.	London.
Meat	None	$32\frac{1}{2}$ pounds.	$31\frac{2}{3}$ pounds.	14 pds. ham.	$7\frac{1}{2}$ pds. ham.	None	None
Salt pork	9 pounds.	13 ... do	$12\frac{1}{2}$ ... do			10 pounds.	
Bread	$13\frac{1}{2}$ .. do	65 ... do	62 ... do	40 pounds.	45 pounds.	15 ... do	$22\frac{1}{2}$ pounds.
Butter		$4\frac{7}{8}$ ... do	$4\frac{3}{4}$ ... do	4 ... do	6 ... do	4 ... do	
Water	60 gallons.	67 gallons.	60 gallons.				60 gallons.
Flour, peas and beans, rice, vegetables	36 pounds.	35 pounds.	44 pounds.	5 pounds.	40 pounds.	40 pounds.	72 pounds.
A further supply of the same articles, in case a corresponding quantity of potatoes may not be had at reasonable prices	$6\frac{1}{4}$ ... do	10 ... do	11 ... do	40 ... do	20 ... do	26 ... do	
Molasses		$1\frac{1}{2}$ ... do	$1\frac{1}{2}$ ... do				$4\frac{1}{2}$ pounds.
Coffee and tea		3 ... do	2 ... do				$1\frac{1}{4}$ ... do
Vinegar	1 pint.	2 quarts.	2 quarts.	$1\frac{3}{4}$ quarts.	$1\frac{3}{4}$ quarts.	2 quarts.	
Sago, wine, sugar, salt, medicines		Sufficient.	Sufficient.	2 pds. salt.	2 pds. salt.	2 pds. salt.	$4\frac{1}{2}$ pds sugar.
Solid food, in Bremen pounds	$64\frac{3}{4}$ pounds.	$155\frac{1}{2}$ pounds.	161 7-10 pds.	$131\frac{1}{3}$ pounds.	$105\frac{1}{2}$ pounds.	90 1-6 pds.	122 6-7 pds.

OBSERVATIONS.

Bremen.—Children above one year of age count for one whole person.
Hamburg.—Children under eight years of age count for one-half person.
Havre.—Children under five years of age do not count.
Antwerp.—Children under eight years of age count for one-half; under twelve years for three-quarters.
Rotterdam.—Children from one to twelve years of age count for two-thirds.
London.—Children under fourteen years of age count for one-half.

But, as the most important difference between the laws of the German States and those of the United States and other countries is to be observed, that at Bremen and Hamburg the ship-owner is in all cases responsible for the prescribed supply of food for each passenger, even when it should have been agreed by contract that the passengers should provide for themselves. This is certainly a most wise and necessary regulation; for, whenever the passengers are left to provide for themselves, as is generally the case in all other ports, experience has too frequently shown that the provisions prove to be wretchedly bad and insufficient. The poor emigrant is but too inclined to rely on the hope that his fellow-passengers will not let him starve, and the consequence is, frequently, that all of them suffer from famine. The United States laws provide that any passenger may, with the consent of the captain, furnish for himself an equivalent for the articles of food required, and that the captain shall furnish comfortable food to such passengers when their own supply, without waste or neglect on their part, should prove insufficient. But this precisely does not protect the emigrant against the consequences of his own want of prudence. He may have supplied himself scantily or with bad provisions on account of his poverty or of his ignorance; and if then his provisions on a long passage come to fail, he is without the means of getting further food, and the captain may even let him die from starvation, as he is not bound to help him, when he can prove that the deficiency was caused by the improvidence of the passenger.

4. Another point on which the German legislation appears to offer more security, is in regard to the control over the vessels that leave with emigrants. No such vessel is allowed to sail, before a public officer has been on board to examine strictly the condition of the vessel, and the supply and quality of the food; the American law also enacts that an inspector of the customs is to examine the vessels, and report whether certain provisions of the act have been complied with; but this examination does not extend to those sections of the act which prescribe the victualling, so that this most important part is under no public control, and the master or owner of the vessel is only answerable to the passengers who shall have been suffering from insufficiency of food. Besides, the German laws oblige the owner or consignee of passenger vessels to insure the whole amount of the passage-money, and, moreover, twenty dollars for each passenger, for the benefit of the passengers, in case the vessel should meet with some accident, and be prevented from reaching the port of destination. These sums are either to be insured at some respectable insurance office, or to be deposited with the government, and are also engaged as a pledge for the compliance with the regulations of the laws. In order to protect the passengers from contagious diseases, it is strictly forbidden at Bremen, as well as at Hamburg, to take on board any person affected with such a disease.

By the above remarks about the laws which regulate the transportation of emigrants at the German ports, it will be seen that the legislation at these places is particularly anxious to provide for the protection and comfort of these people. The beneficial effects of these laws have also been manifested by the generally more healthy condition of vessels arriving from German ports, compared with the frequently very distressed condition of vessels from Belgian, French, and English ports. Among the above stated vessels that arrived recently with so many deaths at New York, only three German vessels occur. Also, the reports of the commissioners of emigration at New York show, that although the German immigration exceeded the Irish in the last year, by far a greater number of Irish fell to the charge of their hospitals—the Irish numbering 6,211, or 5.1 per cent. of the whole amount of Irish immigrants; whereas from Germany only 2,613 patients were admitted, making 2.2 per cent. of the German immigration. This is still a large proportion, but it has to be borne in mind that more than one-fourth of the German emigrants arrive by the way of Liverpool, Havre, Rotterdam, and Antwerp.

No. 12.

Letter from Messrs. Meyer and Stucken, owners of passenger ships.

NEW YORK, *January* 12, 1854.

SIR: In reply to your circular of December 29, we beg leave to submit the following answers to your questions:

I. The space, fourteen feet, allotted to each passenger, is, we think, a full and fair allowance, provided vessels are not allowed to take passengers at the rate of one to thirty feet on second between-deck, or orlop-deck.

II. As to the quantity of provisions required for each passenger, we refer to the Bremen law, § 20; and as to quality, to same law, § 24; said law has proved itself the best existing, and ample in *every* case.

III. Passengers should, in *no* case, be allowed to provide for themselves.

IV. Ventilation. The existing laws, if properly enforced, and the taking of passengers on lower between-deck, (orlop-deck,) is prohibited, are quite sufficient.

V. Cooking arrangements. If the vessel furnishes provisions, all meals should be given at stated hours, and prepared by the ship's cook: any other system is defective.

VI. The laws should give the master more power over the passengers, to enforce cleanliness.

VII. The employment of a surgeon or doctor would be desirable, and would greatly relieve the captain, still it would be unjust and onerous to exact it of every vessel. As it would prove too expensive for small vessels, it might be exacted, properly, with vessels carrying over 300 passengers.

VIII. No attendants are required. We doubt the practicability of the same; and from what our captains tell us, we are inclined to think that it is better to let passengers aid one another. Among 1,500 people there are ever some who, in consideration of small favors from the captain, are not only willing, but glad to be employed, and make themselves useful. A law enforcing a quantity of attendants would be onerous, and productive of new evils.

IX. The separation of sexes, and prevention of unnecessary intercourse between the crew and passengers. The separation of sexes would, we think, not easily be submitted to by families, and if enforced, would not have the desired effect; if illicit intercourse is sought, it can ever be attained on board ship, where so many are crowded together; no law to prevent it could do much good.

X. Process of disinfecting vessel. Under existing laws, where a vessel can carry passengers on several decks, none can be found; the best remedy is, that no vessel is allowed to carry more passengers than will admit of the between-decks being cleared entirely of passengers, cleaned thoroughly, and fumigated with closed hatches.

XI. A report is regularly made by every vessel arriving here with passengers; three copies, one for the custom house, one for the mayor's office, and one for the commissioners of emigration; a fourth report, for the State Department, would be a little extra trouble, and if published, would, we think, have a beneficial influence on captains.

XII. As to inquests being held and verdict published, in cases of death. We hardly think the results would justify the delay and expense incurred; a special form of oath, to be taken by the captain and mate, as to every death, its cause, and that it was not from want or neglect, and such published with list of deaths, would be simpler, and if anything were wrong and worth complaint, it would soon become known.

XIII. As to limitation of passengers in proportion to tonnage. We deem the existing law, 14 feet to each passenger, on first between-deck, quite ample room, provided vessels are prohibited carrying passengers on the lower between-deck, called orlop-deck.

XIV. No distinction in number of passengers for vessels passing the tropics seems necessary, if passengers are not carried on more than one between-deck.

The above are our simple, short answers to your enquiries; but we would beg leave to submit to your consideration a few remarks which may explain them more fully, and show our reasons for them.

We think proper laws in the United States, on the subject of emigration, can do a very great deal toward improving the condition of emigrants on the passage; we fear, however, that legislation on this side of the ocean can, alone, never achieve its object; it should be aided by corresponding laws abroad.

So far as we are able to see, hear, or judge, we find the laws of Bremen the best and most effective. We enclose a copy thereof, which, however, as you require an early reply, we have not had time to translate.

The Belgian law is said to be very strict, but it allows passengers to furnish their own provisions, we believe.

The Hamburg law has been much improved, and remodelled after the plan of the Bremen law.

The British law has been greatly improved, and has one good point in stipulating more boats, or a given number of boats for each vessel.

The French law seems most defective.

Of all vessels arriving with passengers, we believe an impartial report of the commissioners of emigration here will prove that Bremen vessels from Bremen deliver their passengers in the most satisfactory manner; the same with Hamburg vessels from Hamburg and Bremen.

We think it will be found that American and British ships from Bremen deliver passengers more satisfactorily than from any other port, while Bremen vessels from other ports seldom give the same satisfaction as when from Bremen direct. We shall, at first, mention a case of a vessel just arrived from Havre, which will account for it:

A proper report from the commissioners of emigration here will, we think, prove that the per centage of mortality is greatest on large vessels carrying from 600 to 900 passengers; at least we have lately noticed that, while Bremen vessels showed but one, two, and three per cent. mortality, and this mostly infants or aged people, vessels bringing from 600 to 900 passengers showed from eight to ten per cent. mortality. The largest passenger ships have four decks.

1. The upper deck, removable, of strong latting-work, called flush spar deck.

2. The usual deck, caulked water-tight, as every vessel must have it.

3. Lower deck, usual with most vesssels, seldom caulked, and mostly removable, (between-deck.)

4. Lowest deck, or second lower deck, (orlop-deck,) much resembling deck 3, (2d between-deck.)

The deck No. 1 is needed where so many passengers are carried to enable the crew to have room to work the vessel; it affords shelter to deck 2, or regular deck, and leaves that sufficiently ventilated without any other preparation, but it more or less hinders bad air escaping lower decks.

Deck No. 2, or regular deck, is fitted with first class and second class cabins, and the latter is generally large in some; vessels, 200 second class passengers are carried.

Deck No. 3 is the first between decks, with vessels that have no flush-spar deck of latting; is easily well ventilated, but suffers more or less from shelter of spar deck.

Deck No. 4, 2d between-deck, or orlop-deck, is damp, close, and dark, so that a light must be kept burning all day, and its effluvia rise into the upper parts of the vessel. It has another drawback: it is deep down, and people reluctantly climb from it to air and light.

We attribute the greater mortality on board large vessels (taking from 600 to 900 passengers) to their arrangements, and the taking passengers on three decks. The upper deck, or deck No. 2, and first between-deck, or deck No. 3, could be kept healthy, we believe. If no passengers were allowed on deck No. 4, or orlop-deck, (the effluvia arising therefrom, especially in case of infectious complaints, cannot possibly be without disastrous effect to people on the upper decks,) much would be gained. We further attribute the greater mortality to want of proper food and nourishment.

Passengers providing themselves get cheated in every way—quantity, quality, and price. The consequence is, that the poorer lay in not only extreme small stock, but also a defective stock, and trust to the good nature of those better off, or to chances to appropriate to themselves what does not belong to them, and, lastly, rely on the master of the vessel saving them from starvation, but, after all, suffering for want of proper nourishment.

When passengers provide for themselves, they form in sets or companies, each in turn cooking for the set or company. The cook's galley is ever beset; those who have means resort to bribery, those without means resort to force; and serious broils, we believe even murder, have ensued.

It is natural that if a vessel furnishes provisions, the captain or consignee can lay in stores, wholesale, cheaper and of better quality than passengers can in retail; perfect strangers in port of shipment, the captain, owner, or consignee, will, under proper law for liability, see to buy of responsible dealers. The single passenger buys retail, and, unacquainted with sea-voyage, hardly knows what to buy, and buys of unknown and frequently dishonest retailers. The sick and helpless are under this system at a great disadvantage.

In stormy weather, owing to an arrangement of the cook's galley, passengers may, for days together, be unable to cook at all.

If the vessel furnishes stores, and these are measured out in rations to cook, the meals for all passengers can be prepared in two or three kettles, and served at stated hours; the kettles can, as we know they are in Bremen vessels, be so arranged, that even in the worst of weather, warm and properly cooked appropriate meals can be furnished. The defect of improper or deficient nourishment is easily cured.

No sufficient control or guarantee for quality and quantity of stores can be attained, except the same be subjected to rigid inspection before, and at the time of, shipping them.

We believe a law prohibiting the taking of passengers on orlop-deck, enforcing proper ventilation, as now prescribed by law, and making master, owner, and vessel liable for provisions, defining same, their quantity, and that they be sound and healthy, would at once abolish the greatest evils existing. The law should also provide that no ship carries more passengers than will admit of thoroughly cleaning and fumigating between decks.

We have at this moment three Bremen vessels to our address. The captains tell us that, weather permitting, they order at times all passengers on deck; have the between-deck thoroughly cleaned; fumigate with tar, and close down the hatches for an hour, that the fumigation may pierce well. They then air the between-deck, and again fumigate under closed hatches with juniper berries; ventilate afterwards before they allow passengers to go below. Owing to such precautions, we believe, it is that one of these vessels, for which the complement of passengers were waiting at Bremerhaven, and while so waiting on shore, lost fifteen of their number by cholera, had not a single case of serious illness on the whole passage, and no symptoms of cholera. Sea-sickness weakens, makes lazy and indifferent, so that even the most energetic captains complain that they are unable to enforce cleanliness unless aided by law; and threats to treat as mutineers such as will not obey are often necessary. Many passengers have to be brought on deck by main force.

The German society, and reporters for European German papers, collect from emigrants information, and reports are sent to German newspapers stating day of sailing and day of arrival of vessels, with short remarks about weather, treatment, and nourishment on the passage. These are published, and serve emigrants as a guide to what vessels and captains are preferable. Similar publications, by the State Department, would undoubtedly have a good effect. The information could be collected and forwarded by the health officer or commissioner of emigration.

We generally find captains complain most about sickness, &c., on passage, when such has been very rough, keeping passengers below deck and confined. This, and the comparative small mortality on vessels passing tropics, leads us to think that on such voyages no reduction in the number of passengers is required, provided vessels are prohibited carrying passengers on deck four, (second or orlop deck.) The better weather allows more ventilation; passengers are more on deck in open air, take more exercise, and are, in consequence, healthier.

We would finally submit to your consideration the case above referred to.

The Bremen ship, Louisiana, arrived to our care from Havre within the last few days. Her owners, the largest ship-owners in Bremen, never insuring their vessels at all, and employing their vessels mainly between the United States and Bremen, are of course desirous they should bear a good name with emigrants. They consequently take extreme good care to have the best of captains, furnish vessels complete, and have everything in the best order and quality.

They are very severe upon their captains when complaints are made by passengers.

The said ship Louisiana, having gone to Havre with grain, there took a round charter for passengers to New York; vessel to take what American law allows; charterers to furnish stores, medicines, &c.

The vessel has but two decks—the usual upper deck, and one lower deck. The charterer insisted on captain building a second lower deck to enable vessel to carry more passengers. The captain remonstrated, stating his owners would not, even for their own profit, allow it, much less for profit of others, to detriment of passengers. It was proven that all American vessels did it, and he was, under the charter and custom of the port, forced to do for others what his owners never did nor would do for themselves.

As medicines, some castor oil and Epsom salts were handed him; asking for more, he was answered that that was enough, emigrants required no more, &c., and to that effect. Application to United States consul proved unavailing, the consul stating that he knew of no laws which enabled him to interfere.

The captain, knowing his owners, did his duty, and the emigrants have not suffered by the recklessness and hard heartedness of the charterers; but, in one hundred similar cases, not one captain or owner will do what is needed.

Nothing can save the poor emigrant but a general law, making captain of vessel and owner liable for sufficient supply of provisions and medicines, to the extent that they are prevented from chartering vessels to parties who furnish all stores themselves, and inefficiently.

Without such law, resistance will ever be unavailable. Charterers make a profit on stores, and prefer vessels that let them make this profit. The emigrant will be at the mercy of unprincipled, hardhearted speculators, who, living in Europe, cannot be reached by American laws, while captains who do not submit to their terms get no freights.

Steam vessels, whether screw or paddle steamers, we should say, need not and ought not to be as much restricted as sailing vessels. We are, however, unable to give any direct information; we do not, however, doubt that, ere long, steamer lines will be arranged to carry emigrants.

We would respectfully submit for your consideration, that any law entailing increased expense to the ship-owner, while reducing number of passengers in a vessel, would prove more or less onerous, and react disadvantageously on the emigrant.

The profits, especially for *large* vessels, we believe, are such that they can afford to dispense with carrying passengers on orlop-deck, (second between-deck,) and provide satisfactorily for passengers without increasing the price of passage beyond what passage and stores now cost the emigrant.

The profits of smaller vessels, we believe, are hardly of a nature to bear reduction without injury or extra expense to emigrant.

Trusting our remarks and replies to your inquiries may prove of interest, and not be deemed inappropriate,

We are, sir, very respectfully,
Your most obedient servants,
MEYER & STUCKEN.

Hon. HAMILTON FISH, *U. S. Senator,*
Chairman Select Committee on Emigration, Washington, D. C.

No. 13.

Letter from Adolf Rodewald, Esq.

NEW YORK, *January* 20, 1854.

SIR: In answer to your circular, dated December 19, 1853, respecting the causes of "the sickness and mortality prevailing on board the emigrant ships," the undersigned begs leave to state: That the experience gained by him, as merchant, and occasionally consignee of passenger-ships—as member, and, for a term, president, of "The German Society of the city of New York," and, in the latter capacity, for a year, one of the Commissioners of Emigration of this city—has enabled him to form a decided opinion, with regard to the principal causes of the sickness and mortality on board the passenger vessels. And these are—

The insufficiency, or the poor quality, or the imperfect cooking, (and sometimes all combined,) of the provisions.

And it is, further, the decided opinion of the undersigned that the best remedy for these evils would be found in an enactment, making it, in all cases, the duty of the vessel, and the vessel only, to furnish a sufficiency and a good quality of provisions, and also to be responsible for the proper cooking and distribution of the same.

Inexperience, or poverty, or parsimony—and often these combined—

render the passenger quite incompetent to attend to providing the requisite stores. Such a permission is fraught with the most pernicious consequences to himself, and ought, therefore, never to be given.

The very perfect regulations existing at the ports of Hamburg and Bremen to that effect might serve as a guide in framing similar enactments here, and deserve, also, in other respects, the consideration of your committee.

Another main cause of the sickness and mortality among passengers may, no doubt, be traced to the insufficiency of space in the between decks, rendering the supply of pure air, the proper cleanliness, and, under the present state of things, frequently, the access to the cooking arrangements, difficult, and in bad weather, almost impossible.

A sweeping reduction of the present legal number of passengers would, of course, afford a ready remedy for this evil. But your committee will be aware that such a measure would certainly raise the cost of transportation, and, consequently, to a corresponding degree, the price of passage, and would thereby, to a certain extent, check the emigration from Europe, bearing with peculiar hardship upon those classes of the laboring poor in Europe to whom emigration is of the greatest benefit, by keeping them there in their hopelessness and misery. Besides, such a check to the emigration might seriously affect, not only the interest of the ship-owner, but commerce generally, and, in its wider consequences, many important interests connected with the settlement and improvement of the lands in the United States.

The undersigned, in view of these manifold interests, does, therefore, not consider it advisable to reduce the present legal number of passengers per vessel, in a general way, but would rather recommend some amendment to the present law, to prevent an overcrowding in particular instances, by adding to the acts at present in force, the proviso that the number of passengers shall not exceed the former rates of two passengers for five tons, and also some stipulation as to room on deck for air and exercise.

The satisfactory state of things on board of several passenger ships, and especially of many German vessels, proves that, with proper attention to wholesome diet, cleanliness, and ventilation, as well as the use of disinfecting agents, the present number of passengers on board is not incompatible with the maintenance of health.

Many of these preventive means cannot well be regulated by law. The owner, and still more the captain, can only provide for and enforce them; and with carelessness and neglect on their part, the best regulations for these very important matters would fail.

To teach, however, those duties to the careless and negligent ship-owner and captain, by the all-powerful motive of self-interest, the undersigned would beg leave to suggest the propriety of some enactment to the effect that for every passenger dying on the passage, (excepting, perhaps, cases of accident,) the passage money, or—better, perhaps—a certain sum, should be paid over by the consignee to that institution, at the port of arrival, taking charge of the sick and destitute immigrants.

The above views and suggestions embrace the points in your circular from No. I. to VI.

As to No. VII., respecting the employment of a surgeon, the undersigned believes that, notwithstanding all regulations to that effect, the vessels generally would, after all, be provided with inxperienced and incompetent young men, who might do more harm than good, and therefore considers such an enactment inexpedient.

Nos. VIII, IX, X, cannot effectually be regulated by law.

Nos. XI, XII, would seem quite advisable regulations.

No. XIII. The expediency of engrafting this limitation on the present law has been mentioned above.

No. XIV. Such a distinction would, at the first glance, seem advisable—although the undersigned is not able to arrive at any decided opinion on that point—inasmuch as generally, though the heated atmosphere of the tropics may be more detrimental to the health of the passengers, still, on the other hand, the fine weather and smooth sea, more generally prevailing in those latitudes, will allow of better ventilation and more exercise on deck.

In the anxious hope that your committee may arrive at such results, in their report to the Senate, as would induce Congress, in behalf of suffering humanity, soon to pass enactments by which the dreadful state of things on board of many of the passenger vessels may be effectually remedied or prevented, the undersigned has the honor to remain,

With the greatest respect, your very obedient servant,

ADOLF RODEWALD.

Hon. HAMILTON FISH, *U. S. Senate,*
Chairman of Committee on Emigrant Ships.

No. 14.

Letter from Captain William Skiddy.

NEW YORK, *January* 14, 1854.

SIR: In answer to the inquiries in your circular of December 29, 1853, as to the adequacy of the existing laws with respect to passengers, I refer you to the English passenger act of June 30, 1852, which embraces all the inquiries contained in your circular. This act is strictly enforced by the English government, and all American vessels are obliged to conform not only to the act but to many impositions from inspectors appointed by the commissioner; sand we must be subject to this so long as our government permits the law of a foreign country to regulate our ships. After leaving port, however, the master has not sufficient power or authority, either over his crew or passengers, to carry out these regulations.

The *great cause*, therefore, of the filthy, beastly, degrading condition of passengers and their berths, is the want of adequate power in the master to establish a competent police. Under the present navigation laws of the United States, the matter cannot be remedied. The idea of imposing fines on sailors for disobedience and refusing duty, is impracticable. They are almost always in debt to the ship. The same reason or rule will apply to passengers who have scarcely a second shirt.

The master must have more power, or our merchant marine, like the navy, must gradually decline from its former superiority. Even now, I doubt whether one large packet ship could be manned in this port in ten days with American seamen. I regret to acknowledge this, but think it proper the committee should be informed of the fact.

I remain your obedient servant,

WM. SKIDDY.

Hon. H. Fish,
United States Senate, Washington.

No. 15.

Letter from E. D. Hurlbut, Esq.

84 South Street, New York, *February* 1, 1854.

Sir: According to your request, I submit the following for your consideration. With respect to—

Question. 1. The space allotted to each soul on board "three-deck ships," should be as follows: On lower "between-decks," thirty superficial feet; on upper "between-decks," twenty-four superficial feet; in deck-houses and poops, twenty feet; on "two-deck ships," twenty-four superficial feet to each; and in deck-houses and poops, twenty feet.

Question 2. The quantity, kind, and quality of provisions allowed at present is sufficient, when properly taken care of, and properly administered.

Question 3. Permission should not be granted to passengers to furnish or have charge of their own provisions.

Question. 4. The present ventilation is all that is required. (When the number of passengers is limited as in article 1 above, more ventilation would not be required.) The present ventilation often endangers the lives of the passengers and cargo, by being unexpectedly exposed to the breaking in of the sea; side ventilation is and will be more or less controlled and improperly used by passengers occupying such part of the ship, and I consider it decidedly unsafe.

Question 5. The cooking department is sufficient, but badly managed.

Question 6. It should be the duty of the captain to enforce personal cleanliness, and a law is required to protect him in doing so.

Question 7. All vessels should be compelled by law to carry a surgeon, to be recommended or approved by a board of surgeons.

Question 8. Persons employed to administer to the sick and enforce cleanliness, other than the captain and officers and others interested, would be unnecessary, and in most cases would want one or more to take care of them.

Question 9. The separation of the sexes should be in three divisions, as follows: 1st. The married and their children. 2d. Unmarried males. And 3d. Unmarried females. To prevent unnecessary intercourse between the passengers and crew, legislation is necessary to protect the captain in a rigid discipline.

Question 10. A thorough process of disinfecting every vessel where disease has made its appearance is very important, but more necessary to fumigate and keep the vessel cleansed before disease is contracted.

Question 11. The law on this subject is sufficient.

Question 12. Any law on this subject would be violated. Most deaths occur in the severest weather, and in most cases it would be impracticable to hold an inquest.

Question 13. A law regulating the number of passengers by the space allotted to each would be sufficient.

Question 14. All vessels passing between the tropics should be permitted to carry the same number of passengers as those passing north of the tropics.

I would remark, in reference to question 1, that the space allotted as I have named, would limit the number of passengers, so as in most cases to insure health and comfort; and without this limitation, all laws would be ineffective. In reference to question; 2 I would recommend that all provisions, except potatoes, should be packed in tight casks. The present mode of packing bread, meal, &c., is in bags, where from exposure it becomes soured and moulded, and thereby unfit for use. In regard to questions 3 and 5, I would recommend the same law that exists in Bremen, (Germany,) in relation to quantity, kind, and quality of provision, and a suitable number of cooks to be employed by the passenger agents, said cooks to have the entire charge of the provisions, to cook and deal out. In this way the present cooking department would be sufficient, less fuel would be required, and it would add greatly to the health and comfort of the passengers. It frequently happens, that many passengers on board ship are unable to cook for themselves from one week to another, and for the want of properly cooked food become diseased and die. It also frequently happens, that many cook and eat to excess, and, having no exercise, become diseased; this would show the necessity of having a proper cook to prepare the food and deal it out judiciously.

It is very necessary to have the vessels ventilated, fumigated, and cleansed before disease is contracted; in many cases under my notice, cholera, dysentery, and ship fever, have been checked by proceeding to sea. When the weather at the commencement of a voyage has been fair, the ship ventilated and cleaned, discipline and order being established, disease has disappeared; but when the voyage commences with gales of wind, and the officers have as much as they can do to attend to the ship, and this continues a few days, it is then we see, as in recent cases, the effects of sea sickness on previous excess and neglect; and disease makes quick work.

Yours, respectfully,

E. D. HURLBUT.

Hon. HAMILTON FISH,
United States Senate, Washington.

No. 16.

Letter from Dr. Isaac Wood, of New York.

NEW YORK, *January* 10, 1854.

DEAR SIR: I received the circular of your committee a few days since, but pressing engagements have prevented my giving it proper attention. I fear but little value can attach to any suggestion I can make. Permit me to express my gratification that efforts are making to better the condition of those arriving here in our emigrant vessels.

On referring to the laws on the subject, it appears to me not so much is wanting in the laws already existing as in a strict adherence to their several provisions. The space required by law for each passenger appears to me very limited—little more than the superfice of a grave. If, however, due regard be paid to ventilation, it may, in a great measure, do away with the objection. A great evil arises from the passengers occupying this space too much of the passage—in some instances, the whole passage. Being on a committee of the New York Academy of Medicine a few years since, to visit the several hospitals in our city and at quarantine, &c., we visited one of the ships which had just arrived, on board of which thirty or more had died during the passage. There I learned an old man had lain in his berth the whole time since the vessel left port. Whilst on board of this vessel, another one passed us up to the city, on board of which no sickness had made its appearance. I know not what the difference of discipline of these ships was. The one (English) sick, the other (American) well; but I have learned from the emigrants that in some of the vessels the master makes it a rule that all shall go on deck every day, whilst the decks and berths are being cleaned. If any are unable, they are carried on deck. I have an Irishman now living with me who says the captain appointed twenty of the emigrants, directly after sailing, to attend to such duty. In this vessel they had no sickness. Some masters let them do as they please. Hence they remain in their berths until they sicken in their own filth. This might be greatly remedied by requiring every vessel to employ a qualified and experienced surgeon, who shall have a suitable number of attendants to take charge of the sick and enenforce his orders. In this respect there is *lamentable, culpable neglect or deficiency*. Persons are taken on board as doctors who have been nothing but apothecaries, never professed anything else; and some, as I have been assured, having no knowledge of medicine whatever. Why should the owners of ships, who are fattening on the proceeds of these voyages, require the poor emigrant to put his health and life in the hands of one less skilled than their own family physician? In the hands of one they would not trust to treat any domestic animal they valued, is a crying evil.

To show the need of attendants, I would again allude to the visit above named. I saw one family so sick that not one was able to bring the others a cup of water—father, mother, and two or three children. After passing through the vessel, and as I was ascending to the upper deck, I cast my eyes behind, when the emaciated father gave such an imploring look that I could not stand it. I immediately returned to the sick deck, borrowed a cup, threaded my way through the narrow passage to the forward hatch, and obtained a supply of water. On presenting the cup, it was taken with trembling hands by the famished parent, and such was his gratitude that he implored blessings on his benefactor for some time before he would moisten his parched mouth. On inquiring of an Irish girl now living in my family, and who came over in the Manhattan with eight hundred passengers, and was twenty-one weeks on the passage, what this poor family would have done had they been at sea, she replied, if none of the passengers brought them a drink they must do without it. A well educated, regular, and experienced physician and surgeon should be attached to every vessel that crosses the ocean; to every emigrant vessel there should be in addition a suitable number of attendants for the sick.

With respect to the supplies, I think the captain or owners should in all cases be required to provide them. Those required by the law appear scanty but good. The captain providing, the surgeon might be entrusted with directing (under the law) the quality and quantity—one person on ship-board, as on the land, requiring much more than another. The cooking arrangements are too small. The emigrants say the strong make the weak give way until they are served. The appointment of physician and attendants might greatly help this difficulty. To show how very necessary good air and diet are, I would again allude to the official visit to our hospitals, where such immense numbers of fever cases arrived, that to provide for them, several private buildings in different parts of our island were hired, and even tents erected in the open air. I was then informed by the physician to our largest establishment, that those treated in the open air generally did well, or the best, and that the treatment was little more than nutriment. Every morning he had a large quantity of very weak milk punch provided, and gave them liberally to drink. Little or no medicine was required. The poor creatures on ship-board having, perhaps, but a short supply of food of their own, and that frequently of poor quality, and being obliged to cook it themselves or go without, must suffer extremely. Ordinary sea-sickness would disable many during a whole voyage from cooking a meal; hence they become debilitated to such a degree, that when pestilence makes its appearance, they must almost necessarily sicken. In fact, it is almost a wonder that any survive under such circumstances.

With respect to separating the sexes, there can be but one opinion. The frailty of human nature; the bounden duty to protect the weak, the innocent, confiding "weaker vessel," imperatively demands that all practicable guards should be adopted against the demoralizing effects of indiscriminate association of the sexes in such confined quarters as the hold of a ship.

In vessels where pestilence has made its appearance, the ordinary disinfecting agents should be used. My opinion, however, is, that far greater benefit would be derived from turning all on deck, ventilating the hold, washing the persons, clothing, bedding, berths, rooms, &c., than from all other agents which could be used.

I will again mention a fact. About twenty-seven years ago, typhus, ship, or prison-fever, for they are the same, prevailed to an awful degree in our largest prison and pauper establishments, carrying off the officers and physicians in a five-fold proportion to that of the prisoners and paupers. The disease got into the lying-in department, and committed dreadful ravages. Women would be confined, have easy and natural labors, be comfortable twenty-four or forty-eight hours, and the next twenty-four or forty-eight hours be a corpse. The late Dr. Joseph Baily, Joseph M. Smith, and myself, were appointed a committee by our common council, to visit the establishment and advise what means should be adopted to arrest the calamity. We advised the immediate removal of the prisoners, cleansing the building, clothing, bedding, &c., and predicted that in two weeks no new cases would appear. The prediction proved true to the letter. Five only of those which were removed died, and two only from fever alone. They were put in an unfinished building, with large and airy apartments, the doors and windows not yet affixed.

A report of every vessel, as to length of voyage, number of passengers, deaths, &c., would be very useful, as matter for reference for statistical and other purposes, and, I think, ought to be required. In all our prisons a coroner's inquest is required in all cases of death, even where there is a regularly appointed and qualified physician, and I know not why a master of a vessel, and his so-called doctor, should not be as strictly watched. They have quite as much power at sea, and quite as great opportunity for neglect of duty.

I think there should be a limitation of passengers allowed in any vessel, in proportion to the tonnage of the vessel, and also a distinction with respect to the number of passengers between vessels passing within the tropics and those not so passing, but do not feel prepared to express any very definite ideas on the subject. I hope much good may result from your labors on this subject. I fear great abuses have existed under the present system.

Very respectfully, yours,

ISAAC WOOD, M. D.

Hon. HAMILTON FISH,

Chairman of the Select Committee on Passenger Vessels.

P. S.—I embrace this opportunity of expressing to you my grateful feelings for your distinguished attention to the medical profession at the late annual convention of the National Medical Association.

I. W.

No. 17.

Communication from the Board of Health of Philadelphia.

HEALTH OFFICE, PHILADELPHIA, *February* 16, 1854.

SIR: The board of health of this city, have received, through the collector of the port, a copy of your circular letter of December 29, 1853, and have given the subject a careful consideration. It has often, heretofore, been before them, and on this occasion they have, through their committees, aided by the experienced counsels of their solicitor, J. A. Phillips, esq., and of a highly respectable and intelligent shipmaster, long and extensively engaged in the transportation of emigrant passengers, made a thorough examination into our own acts of Congress, and the acts of Parliament of Great Britain, and into the comparative workings of both sets of laws, and they are of opinion, that the British law of 1852, known as the passengers' act, does, more thoroughly than do our own, provide for the due regulation of the subject-matter of the present inquiry, that its provisions are generally more stringent than our own, and that, inasmuch as American vessels engaged in the passenger trade with Great Britain are obliged to obey the laws of both countries, it would be sound policy not only to adopt such provisions of their laws as are improvements on our own, but, so far as practicable, to make arrangements with that and other governments, (by appointing commissioners, or otherwise,) to make all international legislation on the subject conform, as nearly as circumstances will permit.

And, in answer to the several queries in your letter, they beg leave to say:

1. In respect to the allowance of deck room, we do better than the British in this, that we count as adults all persons over one year old, while they count two children, under fourteen years of age, as one adult; and also in this, that part of the allotted space in their vessels on the passenger deck is taken off, fore and aft, for hospitals. Still this board is of opinion that the deck-room is insufficient, and having some time ago carefully examined into the matter, they arrived at the conclusion that the space should be increased from 14 to 20 feet in a vessel 6 feet high between decks, and not passing between the tropics, and should receive a proportionate increase when the height between decks is less, or the vessel passes between the tropics, as in the California ships. The board cannot too earnestly impress this upon the attention of your committee, from its being abundantly manifest that a great proportion of the mortality on ship-board arises from the want of sufficient air and ventilation.

2. In section 4 of the act of Congress of May 17, 1848, there is a provision giving the passenger an option to commute his rations, and this is destructive of the effect intended to be produced by the first clause of the same section. No such option exists in the British law, and there should be none in our own. In Great Britain there is no compulsory allowance of animal food, but a better allowance of provisions otherwise, and their government agents or inspectors always see that they are of the best quality.

3. The master should always be obliged to furnish the provisions, and the passengers should not have the option. The practice of allowing emigrant passengers to find their own provisions is a bad one, and ought to be everywhere abolished.

4. Ventilation is reasonably well provided for by act of Congress, but can be improved by conforming to the British laws, which are more effectual.

5. Here again the British law is better than our own: the size of the caboose, or cooking range, as prescribed by section 3 of our act of Congress, is utterly insufficient for the number of persons therein assigned to it.

6. This is a very important matter, and there ought to be enactments

plainly and clearly defining the duties and powers of the officers in that respect. It is apprehended that section 5 of the act of 1848 is not sufficiently definite, and a recent case in the district court of the United States for this district, where the mate of the ship Saranac was mulcted in damages for a compulsory enforcement on board ship, might lead one to think that the enactments on this subject should be placed beyond doubt or cavil.

7. The employment of a qualified and experienced surgeon should be obligatory.

8. So, also, of the assistants referred to in this query. The British statute provides for the appointment of passengers, stewards, and cooks; and there ought also to be one additional man for every 100 passengers to enforce cleanliness on the decks.

9. This is sufficiently provided for in the British law, and should be incorporated into our own, with this addition, that no one of the crew should be permitted to go below unless by express command or permission of the master or first officer.

10. We have no provision in our law; there is sufficient in the British, which ought to be adopted by us.

11. This is sufficiently provided for by existing laws, and should remain unaltered.

12. It would be impracticable to hold the inquest, by reason of the impossibility of keeping the body, and the presence of the ship's surgeon would render it unnecessary. The existing provisions about reporting the deaths are believed to be sufficient.

13. It is thought that the different modes of computing tonnage here and in Great Britain, and the different models of vessels built here, as regards the qualities of sailing or stowage, would not render a limitation of the passengers in proportion to the tonnage of any practical value. A full allowance of deck room, with due regard to height between decks, will furnish a more reliable guide.

14. Vessels passing between the tropics should allow not less than five feet additional for every passenger.

The foregoing embrace the answers to your specific interrogatories. The board beg leave to commend to your especial attention a very important and interesting publication from the British General Board of Health, entitled "Report on Quarantine, presented to both Houses of Parliament, by command of her Majesty," published in London in 1849, signed by Lords Carlisle and Ashley, and Edwin Chadwick and T. Southwood Smith, in which the above subject is elaborately investigated and discussed, and the suggestions are generally sensible, judicious, and practical. Without going into details, one part of it may be adverted to as being of especial importance and interest.

It appears that, towards the close of the last century, British convicts were transported by contract at so much per head for every one shipped on board. Under this system, the contractors paid no attention to food, cleanliness, or other comforts of the unfortunate passengers, and even over-crowded their already confined space with whatever freight they could obtain. The consequence was, on all occasions, a frightful mortality from typhus fever, caused by filth, foul air, and other privations. "In some of the earlier voyages, full one-half of those who embarked were lost. Later, on the passage to New South Wales, as in the 'Hillsborough,' out of 306 who embarked, 100 were lost; and in another ship, the 'Atlas,' out of 175 embarked, 61 were lost. Yet there were no omissions palpable to common observation, or which could be distinctly proved as matters of crimination, to which responsibility might be attached. The shippers were no doubt honorable men, chargeable with no conscious design against the lives of the human beings committed to their care, and with no unusual omissions; but their thoughts were directed by their interests exclusively, and they saw no reason why convicts or emigrants should not put up with temporary inconveniences to make room for cargo."

This evil was, in the year 1801, effectually remedied, by paying the contractor still *per capita*, but making his pay contingent for such only as were safely landed at their port of destination. The change was almost incredible. By a report from a select committee, it appeared that, from 1795 to 1801, out of 3,833 convicts embarked, 385 died, being one in ten; and after 1801, (when the change was made that put the responsibility on the shippers,) out of 2,398 embarked, only 52 had died, being one in 46; and, subsequently, the average mortality dwindled down to only 1½ per cent., being lower than the average mortality of the same class of persons on shore. The shippers themselves, without any legislative enactments or any supervision, appointed competent medical men to see to the health and comfort of the passengers, and otherwise took such measures as entirely averted the fearful mortality that had before that time more than decimated the unfortunate beings committed to their charge.

The suggestion, therefore, deserves consideration at the hands of your committee, whether this principle might not be advantageously applied to passenger-ships arriving at or departing from any port of the United States, to wit: to make the payment of the passage-money depend upon the arrival of the passenger at his place of destination; and in case of payment being made in advance, and the passenger dying on the voyage, to authorize the surviving husband or wife, parent or child, or other near relative, to recover the money back in admiralty by summary process, death by casualties alone excepted. This would effectually guard the emigrant against the privations and impositions to which he is in many cases subject, in consequence of the competition among those engaged in the business of transporting passengers, (bringing down the price of the passage to the lowest figure,) and would make it the interest of the shipper to take care and keep his passengers in good health, and disarm infectious diseases of their sting.

There is an excellent provision in the British law which we might advantageously adopt, and that is, the necessity of the master giving each passenger his contract ticket, containing a recital of his duties towards the passengers, and the penalties for neglect or dereliction. This should be well printed on a good card, and kept by the passenger. The British form will be found on schedule H of their passenger law of 1852, and is always furnished to those who leave their ports.

By and on behalf of the board of health.

J. G. ELEINA, *President.*
RICHARD GARDINER, *Secretary.*

No. 18.

Letters from A. Schumacker, Esq., President of the German Society of Baltimore, in one of which is a draught of a law, &c.

BALTIMORE, *April* 12, 1854.

I received lately your circular of the 29th of January, as chairman of a committee appointed by the Senate of the United States to inquire into the causes and extent of the sickness and mortality on board of emigrant ships.

It is a subject with which my offices—as president of the German Beneficial Society of Maryland, and consul general of the Hanseatic republics of Bremen and Hamburg—have made me familiar, and I shall be happy to contribute in removing existing evils, and ameliorating the circumstances of my countrymen seeking a new home.

My views regarding amendments and additions to the present passenger laws have already been fully stated in a memorial addressed to the former chairman of your committee, the Hon. Senator Davis, a copy of which I subjoin, and I may therefore limit myself to a partial reply to your several interrogatories:

I. The space allotted to each passenger, of fourteen clear superficial square feet under deck, is, in my judgment, ample. I would recommend, however, that in all cases a certain space should also be left free on the main deck of the vessel—not less than five square feet for each passenger. I believe that many vessels have their decks so much encumbered by cabins, steerages, houses for the crew, &c., that there is actually not room for one half of the passengers to be on deck at the same time.

II. The quantity of provisions required for each passenger by the act approved May 17, 1848, is, in my opinion, not adequate, and I would recommend an assimilation of the regulations in this respect to those adopted at Hamburg, Bremen, and Antwerp, which prescribe about one hundred and fifty pounds of solid food for each passenger. I would also recommend the enactment of a law requiring the captain to have all provisions intended for the use of the passengers inspected, and to bring proof of their having been found of good, sound quality. A certificate to that effect, with a list enumerating the quantity of each article taken on board, and what portion has been left over, should be delivered to the collector at the port of entry.

III. It is decidedly preferable that all provisions should be furnished by the captain, and served in a cooked state. It is unsafe to rely on the prudence of the passengers in providing the necessary stores, and making a proper distribution of them on a voyage of such unequal duration, varying from three weeks to as many months, and frequently prolonged beyond their calculations. Nor is it advisable to let the passengers prepare their own meals, which would cause a continued contest near the cooking apartments, and defeat the ends of equal justice to the weaker parties. Our European emigrants are not yet sufficiently versed in the art of self-government, and will be best cared for when placed under the guardianship of a captain held responsible for their safe and comfortable conveyance.

IV. Ventilation, if solely dependent on the two ventilators prescribed in the law approved May 17, 1848, is imperfect. All vessels should be provided with ports in the stern or cabin, to be kept open when the weather will in any way permit it. Small ports in the side of the ship are also valuable auxiliaries for keeping the air pure.

VI. It is not expedient to make the captain responsible for the personal cleanliness of the passengers, and he should only be held accountable for the cleanliness of the vessel.

VII. The employment of a surgeon is of questionable utility. I doubt whether properly qualified persons could be procured for several hundred emigrant ships; and it is to be apprehended that, in a majority of cases, the place would be filled by young inexperienced hands. I believe that it will be safer to leave the administration of all medical and surgical advice and aid to the captain, who has always more or less experience in the treatment of accidents and maladies incident to sea voyages, and who is thus generally more competent to give relief to his passengers, and even in cases of emergency, than young surgeons and physicians who have never been on salt water. The law would moreover give an undue advantage to the largest class of ships, and go far towards excluding small vessels from participating occasionally in the passenger business, as the expense of engaging a surgeon for a single voyage would be too onerous.

VIII. It is unnecessary to pass special laws in this respect. Nurses and attendants can always be found among the passengers if wanted; and it would be injurious to compel their being previously engaged at the port of embarkation at an additional and perhaps unnecessary expense.

IX. Arrangements for the separation of the sexes will be difficult, and attended with inconveniences. It might be tried, as proposed in my former memorial, by berthing all the unmarried male passengers over a certain age in the forepart of the ship, but this would frequently interfere with the ventilation of the vessel, and I am not prepared to pass a decided opinion as to the expediency of the measure. I would not advocate special laws to prevent intercourse between the crew and the passengers, as it would be next to impossible to maintain them, and the police of the ship ought, in my opinion, to be left to the officers.

XI. A report as proposed may be useful, and would not give much trouble.

XII. An inquest to be held subsequent to the arrival in port is, in my opinion, not important, and would cause an expenditure of time and money, which I should be unwilling to incur.

XIII. A limitation of the number of passengers in proportion to the tonnage, I consider highly desirable beyond any other reform. I would recommend a restoration of the former law, fixing the number of passengers allowed to be carried to two for five register tons, (children below the age of one year not to be counted;) and, also, that only a small portion of the passengers be permitted to be lodged otherwise than in the apartment immediately below the main deck. In fact it would be advisable to prohibit altogether the carrying of passengers on the lower, or orlop-deck, in vessels having three decks.

XVI. A distinction between the number of passengers allowed to vessels passing within the tropics, and those not so passing, is not deemed requisite; and any restriction should, at all events, be confined to the summer months.

I would further recommend that the law regarding the construction of separate berths for each passenger be repealed. It is too inconvenient to be enforced, and I have good reason for believing that, in many instances, if not generally, the partitions between the berths are taken down by the passengers almost immediately after coming on board, and only put up again when the ship is near entering her port of destination.

The English passenger laws prescribe that each passenger ship shall be provided with a certain number of boats, life-boats, life-buoys, privies, &c., and that a separate compartment be assigned and fitted up as a hospital. I do not wish to dispute the usefulness, to a certain extent, of these regulations, but would remark that care must be taken not to enhance the price of passage by too many laws. I believe that, as respects the German ports, the manner and extent of the outfit may be left to the authorities there, so long as the latter will continue to exercise the same strict supervision as heretofore. I am not aware that complaints have ever been made here of passengers having been put on short allowance, even after unusually protracted voyages. On the contrary, I believe that too free an indulgence in the substantial and plentiful fare offered on board, and to which the emigrants from the interior of Germany are not accustomed, is frequently a cause of sickness.

The most effectual measure for insuring good treatment and close attention to the health of the passengers would, perhaps, be a law empowering the visiting physician at the port of entry to send any sick passenger, so willing, to one of the city infirmaries, there to be boarded and cured at the vessel's expense.

A law prohibiting passenger ships from carrying gunpowder, vitriol, guano, green hides, or any other article liable to endanger the health or lives of the passengers, might also be of service.

All the regulations stated above, or others in furtherance of the same object, should only be applicable to passenger ships, or vessels carrying beyond a certain number of passengers; and I would recommend that a limited number, say 25 or 30 passengers, should be allowed to be transported without restriction of any kind.

The penalties for transgressions of the several passenger laws should be revised. I think that the substitution of smaller fines, to be rigidly exacted, would better attain the object sought after than the present heavy penalties, amounting in some cases to confiscation of the vessel, as the very severity of the law often prevents its being made operative.

I hold myself in readiness to give you any further information in my power, and avail myself of this opportunity to assure you of my regard.

A. SHUMACKER,
President of the German Society in Maryland.

Hon. HAMILTON FISH,
Chairman of Committee on Commerce, Washington City.

BALTIMORE, *April* 8, 1847.

SIR: The attention of the undersigned has been directed to an act of the last Congress, entitled "An act to regulate the carriage of passengers in merchant vessels," and to your circular addressed to collectors and other officers of the customs, giving directions regarding its strict observance, &c.

This act, though manifestly intended for the benefit of the passengers, is believed by us to contain several objectionable features, calculated to prove injurious, and to check the emigration to this country. The old law, now in force, restricts the number of passengers which each vessel is allowed to take to two for five tons, register measurement, and we believe that ample accommodations can be furnished to that number, particularly by larger vessels having their cabins on the upper-deck, such as are now generally employed in the trade with Europe. Our character as officers of the German Beneficial Society and consuls general, affords us a good opportunity to judge of this matter, as it brings us into daily contact with our newly arrived countrymen, asking our advice and assistance, and we are not aware that complaints have ever been made to us about want of room on board a ship. The above act, changing the law so as to assign a certain number of superficial square feet to each passenger, has its orgin in an abuse by which the space intended to be set aside for the exclusive use of the passengers had partly been appropriated to other purposes, and filled with stores and merchandise, thus confining the passengers to a much narrower space than the law designed to give them. This has chiefly occurred on board of vessels arriving at New York from Liverpool, whence large quantities of bulky articles, such as salt, coal, and iron, are usually shipped, while the goods exported from Bremen, Hamburg, and Antwerp, are less voluminous. It is a rare circumstance that more freight is offered at the last named ports than can conveniently be stored in the lower hold, and there is thus no inducement to encroach on the room between deck which ought to be allotted to the passengers. Fourteen clear superficial square feet on the lower-deck, or platform, for each passenger, is, in our opinion, an over-ample allowance, and will reduce the number which each vessel is permitted to take under the old law by at least one-third, in some cases by one-half. We doubt whether the framers of the new law have contemplated so great a reduction, and it would be quite inappropriate to go yet further and require an additional space for the berths. Though the act admits of that interpretation, we have construed it differently, viz: that the space occupied by the berths shall be included in the fourteen clear superficial feet allotted to each passenger, and we are confident that our views are shared by the public in general and ship-owners in Europe. The latter will have acted in accordance with their construction of the act, and, if yours should be adhered to, all vessels departing prior to the promulgation of your circular, and arriving after the 31st of May next, will incur the penalty of forfeiture. The last named date, when the law is to take effect here, is even too near for a sufficient promulgation of the original act as approved on the second of March. The latter has probably not reached the north of Europe till early this month, perhaps not before the 15th, and vessels leaving the

Elbe or Weser at that time, are not considered due here before the end of May or early in June, sixty days being an ordinary passage. We should, therefore, strongly recommend that the time when the law is to go into operation, as regards vessels from the north of Europe, be fixed at least one month later.

The regulation prescribing the dimensions of the berths is unsuitable in our opinion. Six feet by eighteen inches is too little for a grown person, and inconveniently large for children. As the ship does not furnish bedding for steerage passengers, the latter would be put to an extra expense by having matrasses expressly made to suit the narrow size of the berths, and to obviate this, many would, no doubt, prefer sleeping on their own bedding spread on the floor of the ship. The law, if carried into effect to the letter, would be absolutely onerous for passengers wishing to leave here for Europe. Room is more valuable on that voyage than to this country, and the price of passage would be enhanced accordingly. If there is a ground to protect emigrants from Europe by laws conducive to their safety and health, as a class of people totally unacquainted with the sea, a majority of whom have never seen a large vessel, we can perceive none which should induce this government to become their guardian, and interfere with their liberties, when wishing to return to the land of their ancestors. The voyage is generally made in the summer season; if a dozen or more passengers offer, a hut is frequently erected on deck; a smaller number, say four or six, can usually be lodged in the large boat, where they are even placed more comfortably than below, and if this arrangement suits the convenience of all parties, government ought not to interfere. In numerous instances, our German captains have been induced to take persons wishing to return for a trifling compensation. The German Society has often paid the passage for individuals, who would otherwise become a charge to the public; we could even cite cases where entire families have been sent back, all which cannot be done in future if the regulations of the new passage-law are rigidly enforced.

We would also respectfully bring to your notice, that if a difference is to be made as regards vessels passing within the tropics, limiting them to a smaller number of passengers, such restrictions should be confined to the summer months. The great body of the emigrants annually going to New Orleans and Galveston, leave Europe late in fall or very early in spring, at a season when they are even less exposed to the heat of the sun than their countrymen arriving at our northern ports in July and August. Vessels bound to the Mississippi and Gulf of Mexico generally go south of Cuba, thus entering the tropics. The voyage over the Bahama banks, though shorter, is more perilous, and it would be injudicious to oblige captains to steer that course in order to avoid the penalty of the law.

In conclusion we would beg leave to state, that the transportation of passengers from the Elbe and the Weser is under the direct supervision of the authorities there, who permit no vessels to take emigrants till after rigid inspection of the arragement for the accommodations, stores, etc. It is also clearly the policy of ship-owners to see all passengers treated well, and thus establish a good reputation for their vessels and captains, who are always enjoined to give satisfaction.

If any modification of the passenger laws should be concluded upon, we would thank you to give us the earliest information for transmission to Germany.

A. SCHUMACHER,
President of the German Society, &c.,
T. L. BRAUNS,
Vice President of the German Society, &c.

Hon R. J. Walker,
Secretary of the Treasury, Washington.

Whereas, the existing laws to regulate the carriage of passengers in merchant vessels have proved defective and inadequate in certain respects, be it therefore enacted by the Senate and House of Representatives of the United States of America in Congress assembled, that the following amendments and additional regulations shall take effect from the ——, and that so much of the acts approved February 22, 1847, and May 17, 1848, as are not in conformity with the present act, are hereby repealed.

Sec. 1. No master of any vessel owned in whole, or in part, by a citizen of the United States of America, or by a citizen of any foreign country, shall take on board such vessel, at any foreign port or place, with intent of carrying the same to a port in the United States, a greater number of passengers that in the proportion of two passengers to every five tons of the registered tonnage of such ship, of which number not more than in the proportion of one-fourth, and not exceeding one hundred and fifty passengers, shall be placed or lodged otherwise than on the deck or platform immediately below the main-deck of such ship. It shall in all cases be computed that two children, each being under the age of eight years, shall be equal to one passenger, and that children under the age of one year shall not be included in the computation of the number of passengers. All laws and regulations, in regard to the clear unencumbered space to be allotted to each passenger, shall remain in force.

Sec. 2. In every passenger ship the beams supporting the passenger decks shall form part of the permanent structure of the ship. They shall be of adequate strength, and shall be firmly secured to the ship. The passenger decks shall be at least one inch and a half in thickness, and shall be laid and firmly fastened upon the beams continuously from side to side of the compartment in which the passengers are berthed, or substantially secured to the beams at least six inches clear above the bottom thereof.

Sec. 3. In every passenger ship all the unmarried male passengers, of the age of fourteen years and upwards, shall be berthed in the forepart of the ship in a compartment divided off from the space appropriated to the other passengers by a substantial and well-secured bulkhead, or in separate rooms, if the ship be divided into compartments and fitted with enclosed berths. The berths shall be securely constructed, and of dimensions of not less than after the rate of six feet in length and eighteen inches in width, for each adult passenger. Not more than two passengers, unless members of the same family, shall be placed in the same berth, nor in case shall persons of different sexes above the age of fourteen, unless husband and wife, be permitted to occupy the same berth. No berths in a passenger ship, occupied by passengers during the voyage, shall be taken down until forty-eight hours after the arrival of such ship at the port of final discharge, unless all the passengers shall have voluntarily quitted the ship before the expiration of that time. Not less than two lanterns shall be kept burning in the between decks from sunset to sunrise.

SEC. 4. In every passenger ship a space shall be properly divided off and set apart for an hospital, not less, in ships carrying as many as one hundred passengers, than fifty-six clear superficial feet, with four bed-berths erected therein, and properly supplied with bedding; nor less, in vessels carrying three hundred or more passengers, than one hundred and twelve clear superfical feet, with at least six bed-berths, supplied as aforesaid.

SEC. 5. Every passenger ship shall be provided with at least two privies for the use of the passengers, and with two additional privies for every one hundred and fifty passengers on board, the entire number not required to exceed ten; such privies shall be placed on each side of the ship, and be kept in a serviceable condition throughout the voyage.

SEC. 6. Every passenger ship shall carry a number of boats in a sea-worthy condition, according to the following scale, that is to say: two boats for every ship of one hundred tons and upwards; three boats for every ship of two hundred tons and upwards, if the number of passengers shall exceed fifty; four boats for every ship of five hundred tons and upwards, if the number of passengers shall exceed two hundred; five boats for every ship of eight hundred tons and upwards, if the number of passengers shall exceed three hundred; six boats for every ship of twelve hundred tons and upwards, if the number of passengers shall exceed four hundred and fifty.

One of such boats shall, in all cases, be a long-boat, and one shall be a properly fitted life-boat, which shall be kept properly suspended at the quarter or stern of the ship; and each of such boats shall be of a suitable size, properly supplied with all requisites, and kept clear at all times for immediate use at sea. There shall likewise be on board of each passenger ship two properly fitted life-buoys, kept ready at all times for immediate use, and also a fire-engine in proper working order, or other apparatus for extinguishing fire.

SEC. 7. No passenger ship shall have on board as cargo, gunpowder, vitriol, guano, green hides, or any other article, whether as cargo or ballast, which by reason of its nature or quantity, shall be likely to endanger the health or lives of the passengers or the safety of the ship.

SEC. 8. Each passenger ship shall have on board, for the use of the passengers, at the time of leaving the last port, well secured under deck for each passenger, at least twenty pounds of salt beef; ten pounds of salt pork, both to be free of bone and of good quality; thirty-five pounds of bread or biscuit, equal to good navy bread; ten pounds of rice; ten pounds of oatmeal; ten pounds of wheat flour; ten pounds of peas and beans; thirty-five pounds of potatoes; one quart of vinegar; sixty gallons of fresh water, and a sufficient supply of fuel for cooking. At places where either rice, oatmeal, wheat flour, or peas and beans cannot be procured of good quality and on reasonable terms, the quantity of either or any of the other last named articles may be increased and substituted therefor, and in case potatoes cannot be procured on reasonable terms, one pound of either of said articles may be substituted in lieu of five pounds of potatoes. The captains of such vessels shall deliver to each passenger about one-tenth part of the aforesaid provisions weekly, commencing from the day of sailing, and daily at least three quarts of water, which latter shall be carried in tanks or in casks, to be sweet and tight, of a sufficient strength, and properly charred inside, not made of fir or soft wood staves, nor capable of severally containing more than three hundred gallons each. He shall also deliver sufficient fuel for cooking, and if the passengers on board of any such vessel in which the provisions, water, and fuel herein required as aforesaid, shall not have been provided as aforesaid, shall at any time be put on short allowance, the master or owner of any such vessel shall pay to each and every passenger the sum of three dollars for each and every day he or they may have been on such short allowance, to be recovered in the circuit or district courts of the United States: provided, nevertheless, that nothing herein contained shall prevent any passenger, with the consent of the captain, from furnishing for himself the articles of food herein specified; and if put on board in good order, it shall fully satisfy the provisions of this act as far as regards food; and provided further, that any passenger may also, with the consent of the captain, furnish for himself an equivalent for the articles of food required in other and different articles, and if without waste or neglect on the part of the passenger, or an inevitable accident they prove insufficient, and the captain shall furnish comfortable food to such passengers during the residue of the voyage, this in regard to food, shall also be a compliance with the terms of this act. But in all cases the provisions so furnished shall be under the sole control and in keeping of the captain, who shall be responsible in the same manner as if all articles had been procured by him.

Sec. 9. The owner or charterer of any passenger ship shall provide for the use of the passengers a medicine chest, containing a supply of medicines, instruments, and other things proper and necessary for diseases and accidents incident to sea-voyages, and for the medical treatment of the passengers during the voyage, including an adequate supply of disinfecting fluid or agent, together with printed or written directions for the use of the same respectively. No charge shall be made to any passenger for such medical or surgical treatment.

Sec. 10. No person shall in any passenger ship during the voyage directly or indirectly sell, or cause to be sold, any spirits or strong waters, on pain of being fined for every such offence not more than one hundred dollars nor less than twenty five dollars.

Sec. 11. Any transgression of the several laws and regulations contained in sections 2, 3, 4, 5, 6, 7, and 9, shall subject the captain or owner of the vessel to a fine of five hundred dollars.

BALTIMORE, *July* 22, 1852.

In accordance with your permission, I hand you herewith a statement embodying my views of the changes and additions which it would be desirable to be made in the existing passenger laws now under revision. The three principal causes of sickness and death are, crowded ships, want of cleanliness, and inappropriate food, and it becomes the duty of all friends of humanity to devise measures for the removal of these evils. The act to regulate the carriage of passengers in merchant vessels, approved February 22, 1847, had this object in view by allotting a fixed, unencumbered space to each passenger, but its benefit was in a great measure lost by the repeal of the act approved March 2, 1819, which made the number of passengers each vessel was permitted to carry dependant upon her tonnage, and limited the same to two passengers for every five tons register. The abolishment of this restriction has led to the erection of extra cabins and compartments, by which vessels are now entitled to carry even a much larger number of passengers than under the old law, and instances could be cited that ships of a particular construction, measuring less than 1,200 tons, are permitted to take over 800 passengers, or nearly double the number than under the old act. Though ways and means have thus been found to accommodate such a mass of human beings at night, assigning to each person a space of 14 clear superficial feet and a separate berth, no proportionate room can be given to them during the daytime. On the contrary, the space on the main deck is to such an extent encroached upon by the additional cabins and houses, that frequently not one-fourth of the number of passengers on board can be on the upper deck at one time without being in the way of the crew, and the rest of the passengers are compelled to remain in the foul atmosphere below. This circumstance renders it also extremely difficult to preserve the requisite cleanliness in the always crowded between deck, and has no doubt often engendered diseases. Additional regulations, regarding the quantities of provisions and water which each passenger ship shall carry, are almost superfluous, as the laws and regulations on that head, in all the principal ports of Europe, are very explicit, and prescribe a much larger stock of stores than the laws of the United States; in proof of which, I beg to refer you to the enclosed list of outfits at Bremen, Hamburg, Havre, Antwerp, Rotterdam and London, contrasted with the scanty list prescribed by the act of May 17, 1848. No instance has ever come under my cognizance of passenger ships arriving from Germany short of provisions, even after unusually long passages. On the other hand, I consider it of vital importance, that the law should enact an obligation on the captain to issue all provisions in a cooked state, and to prohibit all passengers from being the keepers of such provisions as they may bring with them. The want of warm, nourishing food has, in many cases, been destructive to the health of the passengers, and it is dangerous to rely on their prudence and judgment in the preservation and proper division of their stock of provision on a voyage sometimes extending over two months. Other additions to the existing passenger laws might, perhaps, be suggested, but I am almost afraid that too much legislation would prove injurious. The last English passenger law is entirely too complicated and too one-sided, and if rigidly enforced it must check emigration from Great Britain.

If you want any other information on the subject, which I can give you, I shall do so with pleasure, and meantime I remain, with true regard,

Your obedient servant,

A. SCHUMACHER,

President of the German Beneficial Society in Maryland.

Hon. SENATOR DAVIS,

Washington, D. C.

No. 19.

Report of the Board of Trade of Baltimore.

OFFICE OF THE BOARD OF TRADE,
Baltimore, March 13, 1854.

DEAR SIR: At the last regular monthly meeting of this board, held on the 6th instant, the annexed report was adopted, and a copy ordered to be forwarded to you.

Very respectfully, your obedient servant,

GEO. U. PORTER,
Secretary.

Hon. HAMILTON FISH,
Chairman, &c., U. S. Senate.

The Committee on Commerce of the Board of Trade of Baltimore respectfully report:

That they have carefully considered the questions proposed by the Honorable Hamilton Fish, chairman of the committee of the Senate of the United States on the subject of sickness, &c., on board of emigrant ships.

That they coincide generally with the answers already given to the questions in full, by other parties from this city, and consider it unnecessary, therefore, to reiterate replies in detail.

That in the judgment of this committee the laws now in force, with the additional provisions mentioned below, are sufficient for the protection of passengers on board of emigrant ships.

First. That no vessel of three decks be allowed to transport passengers on the orlop or lowest deck.

Second. That at least one hundred and fifty pounds of solid provisions should be required to be placed on board for each passenger, on a voyage from Europe to the United States, and a proportionate quantity for other voyages.

Third. That all provisions should be required to be furnished and cooked by the vessel.

Fourth. That the fines for violations of the passenger laws should be more moderate and rigidly enforced.

Fifth. That the laws on this subject, of this country and those from which the emigrants sail, should be assimilated, as the diverse requirements of the laws of various countries interpose obstacles to free emigration.

THOS. WHITRIDGE,
Chairman of the Committee.

No. 20.

Communication of Dr. Cartwright, of New Orleans.

Some general remarks on the subjects embraced in the circular issued by a select committee of the United States Senate, (Hon. Hamilton Fish, chairman,) inquiring into the propriety of further legislation in regard to emigrant ships, &c., by SAMUEL CARTWRIGHT, M. D., *New Orleans.*

CANAL STREET, NEW ORLEANS,
March 4, 1854.

The greater number of emigrants are in humble life. Few of them know anything of ships, or of the precautions necessary to insure a safe and comfortable passage. They seldom can gain the requisite information in regard to the fitness of a ship for the conveyance of passengers, not knowing whom to apply to among the disinterested possessing the requisite knowledge.

Hence, unprincipled masters and owners are not slow to take advantage of their ignorance, by offering them a cheap passage, enticing them on board of vessels rejected by the insurance companies as unseaworthy, and unfit for carrying cargoes of valuable goods.

The destruction of life by shipwrecks and disease has been most appalling among the emigrants who have been enticed on board the worn-out vessels engaged in the Canadian timber trade—seventeen being shipwrecked in a single season in the Gulf of St. Lawrence, and more than 700 lives lost.

The voyage to Quebec—owing to the contrary winds and rough weather in that inhospitable climate—is apt to be much longer than to any other part of America. So safe, agreeable, and expeditious is a voyage from Europe to the southern portion of this republic, in autumn and the beginning of winter, that the wealthy natives, in returning home from Paris and other places in Europe—which many of them frequently visit—generally prefer taking passage on board the safe and substantial No. 1 ships annually visiting our southern ports for cargoes of cotton.

The voyage, so far from being dangerous or sickly, is most agreeable and healthy. Yet emigrants, to save a few dollars, not knowing that our southern ports are always free from yellow fever and other malignant diseases in November and December, when the cotton ships arrive, are apt to be enticed on board the worn-out Canadian timber ships, where storms, disease, and all the discomforts of an angry ocean await them.

So full of danger is a voyage from Europe to Canada, even in the first-class vessels, that the wealthy and cautious English prefer reaching the British provinces by the way of New York or Boston, as not only safer, but more expeditious. Policy and humanity both require that the price of passage should be made as cheap as possible to the emigrants, consistent with their health, comfort, and security. An unnecessary quantity of expensive provisions, the employment of a surgeon on board of every passage ship, with regular nurses, cooks, and attendants, and too strict a limitation of the numbers admitted to embark, would be objected to by passengers and shippers, as involving so much expense as to prevent a very large portion of worthy people from ever reaching these free shores.

On the other hand, the crowding of too many together, or the want of a sufficient supply of food and water of good quality, is attended with such serious consequences as imperiously to require additional legislative regulations to prevent their recurrence. Legislation could do much good in prohibiting all ships not perfectly sea-worthy from carrying any passengers; also, all those less than 5½ feet between decks. It is believed that the price of a passage in a sea-worthy vessel, with a plenty of water and food of good quality, and sufficient space for each emigrant, need not be raised but little, if any, above the present rates; provided that wholesome rules and regulations be prescribed by law to insure order, comfort, and economy.

The first on the list to be, the absolute prohibition of ardent spirits or fermented drinks in any form, either among the passengers or crew of any ship authorized to carry emigrants. Large numbers of American ships are now, and have, for many years, been sent to sea on long voyages, on board of which ardent spirits have been strictly prohibited. Where coffee and cocoa are dealt to the crew in place of the usual grog allowance, there is so much more safety, that life and property are insured on easier terms than on any other vessels. This fact outweighs all objections heretofore made against the disuse of grog. When seamen are employed on a ship, they sign an agreement to conduct themselves in an orderly, faithful, and careful manner, and to be at all times diligent in their respective duties and stations; also, to be obedient to the lawful commands of the master in every thing relating to the ship, in the line of their respective duties and stations.

The master has no right to order the cook (who in technical language is a seaman) aloft to furl the sails, or to put the mate to scrubbing; because it would be out of the line of the prescribed duties and stations of each. It is only when the ship is in peril that the master becomes an unlimited despot; he can then order anything to be done by any person on board, within the line of his duty or not. He can even compel the passengers, whether in the cabin or the steerage, to toil by night or by day at any species of drudgery work to secure the general safety.

A law requiring, as a preliminary, all emigrant passengers to sign a written agreement, binding themselves to behave in an orderly and becoming manner, and to be obedient to the lawful commands of the master in all things appertaining to their duties and stations as passengers, would be most desirable, if coupled with a provision, distinctly enumerating and defining the duties appertaining to the several classes of passengers. This would be much better than the present law, requiring masters to post the rules of the ship in some conspicuous part of the vessel. It would be better for Congress to make the rules; and among those rules should be certain hours devoted to repose and to particular exercises, viz: dusting their wearing apparel; washing their persons and dry-rubbing their several apartments; airing their bed-clothing, and cleansing their hammocks or berths, and the surrounding walls from all impurities. Twice a week a general washing of all soiled garments and bed-clothing admitting of no other method of purification. Also, to act as nurses for the sick; to assist in cooking for them when required to do so; to take medicine, when sick, from the hands of a surgeon, if one be on board; and, moreover, to permit their children, during certain hours in the day, to be under the control and authority of any person that might be selected, who was willing and competent to give them instruction.

It is believed that the school-master abroad on the sea could be of infinite service to the emigrants, both to adults and children. Even without being familiar with their language, he could impart much useful information to them, by the black-board and maps; giving them useful lessons regarding the extent and geography of the great republic so soon to be the home of themselves and their children forever. Many young physicians, for the sake of visiting Europe and seeing the world, would, no doubt, be willing to act in the double capacity of surgeon and instructor to the emigrants on any good ship giving them a free passage to Europe and back, or for a small compensation.

There are questions constantly arising, in this great and growing country, which the wisest legislators the world ever produced find much difficulty in meeting, when unaided by those who have made the specialities with which they are often connected the subject of their particular study.

Much of the British legislation on new questions oftener comes, in its incipient stage, from the humble studio of the savan than from the palace of the earl.

The new questions springing up at every step in the progress of the British empire over the eastern continent, have been and still are referred to unpretending scientific men, to gather all the facts, and to make a full investigation of everything connected with them, as a preparatory process to legislation.

In addition to similar questions, growing out of the expansion of our republican empire westward, others have arisen in regard to the best legislative measures to promote the safety, health and comfort of the noblest varieties of the human race, the Celts and the Teutons, in their transit from the despotisms of Europe to the beloved land of their dreams, where the people in peace and plenty reign.

God has made the ocean healthy, and if it prove otherwise, it is in consequence of man's tyranny, ignorance, or avarice, pursuing the emigrants on their passage hither, requiring the interposition of republican legislation to correct the evil. The legislation, however, to effect so desirable an object, should be founded on a knowledge of the causes endangering life and producing disease and discomfort at sea.

The peculiar knowledge of the several professions, as well as that taught by the experience and observation of practical men, should all be brought into requisition. But more especially the discoveries of those who, like Dowler and Avequin, of this city, have spent the most of their lives in gathering materials for enriching science by following nature into every hidden retreat, should not be permitted to remain unknown in musty manuscripts, when their publication would throw much light on many subjects connected with the new questions which our country's rapid growth is bringing forward for legislative action. Avequin, for instance, on the sugar question.

Pure science, pursued for the love of science, does not pay; and newly discovered truths, conflicting with the errors and prejudices of the day, are generally too unpopular to bear publication, except at the expense of those who have spent their lives and wasted their means in their discovery.

Such, for instance, as the truth announced by Dowler, and demonstrated from the geological formation of the soil, (containing five separate cyprus forests, each superimposed above the other,) that New Orleans stands on ground upwards of one hundred thousand years old, and also other facts proving that it is one of the healthiest locations for a great city in the world. So far are the facts establishing the great age of the world from standing in conflict with the truths of the Bible, as some suppose, they rather serve to restore the true sense of certain Hebrew words, of which we have lost the meaning. As the word translated "*days,*" for instance, evidently means vast cycles of time.

Natural history is a revelation of God's works and will, and is well calculated to correct the verbal errors that have crept into the Sacred Writings, and to restore their true reading.

Great Britain rewards her scientific men with what our government very properly regards as empty honors, and has prohibited by law.

Titles of nobility, as the prefix of the word "sir" to the name of those who have enriched science with valuable discoveries, however empty and illegal here, are valuable there in leading to patronage and privileges which could not otherwise be enjoyed. Our patent laws only apply as means of encouragement to the arts and subjects connected with mechanical inventions.

The encouragement thus held out covers only a part of the field occupied by our men of genius. But as far as it extends it has done wonders. It has given us the steamboat and the electric telegraph, and many other things contributing to place the United States above all nations in useful inventions. If behind other nations in the higher walks of science, it is because our scientific men have no proper encouragement, either from the people or the government, and every step they take in purely scientific investigations is a step to poverty, and acts as a disqualifying circumstance in gaining a livelihood or obtaining any office in the gift of the government.

Those occupied in laborious scientific investigations have no time or inclination to electioneer and to seek for patronage. The great chemists, Liebig and Dumas, quote Avequin's discoveries in regard to Louisiana sugar. But he is old and poor—almost unknown in New Orleans, where he has resided for a third of a century.

Dowler was unanimously recommended by the city council of New Orleans, as eminently qualified for almost any office in the gift of the federal government. Yet he got none. And when vacancies occurred in the gift of the city council, the very council which unanimously recommended him, some one always steped in before him and got every lucrative office while he was at his studies.

But for this neglect of men of original genius in the higher walks of science the country would not be deprived of their services on the springing up of new questions, requiring not only more learning and experience, but an original mode of thinking, to combine their new relations, and to disentangle them from old errors.

One of the greatest nuisances of a political life consists in the interminable swarms of importunate, and often undeserving, office-seekers, besetting the pathways of every politician or person in authority, who has any office in his gift, or influence with those who have it to bestow. A provision rigidly observed of giving the preference in appointments to office, other circumstances being nearly equal, to those only who in arts, in arms, or in science have conferred some public benefit on their country, it is believed, would not only abate the above-mentioned nuisance, but do more in kindling a spirit of invincibility in war, and in arousing the energies of men of genius in peace, than the British custom of conferring orders of nobility.

American legislators have not the same advantage that the English and French have in being able to command the individual knowledge of all the savans of the land on any matter connected with their duties as legislators. They must possess the knowledge within themselves, or legislate in the dark.

No one body of men can know everything. The subjects embraced in the circular emanating from the committee of the United States Senate, inquiring into the causes of the sickness and mortality prevailing on board the emigrant ships, and the insufficiency of existing laws, cover a wide field, much of which none but the medical profession has ever explored or can explore, and they not thoroughly, having themselves to depend for much of their knowledge on the scattered observations and manuscripts of those devotees who pursue science not for gain, but for love.

That the special devotees of science—those laboring for the public good, and the politicians of the same order—should be the only classes of people whom hard labor brings to want in this country is a defect in our institutions calling for some appropriate remedy.

A remedy for that defect would not only be the surest way to gather in all the information that can be had, or that can be extorted from nature on the subjects embraced in the honorable committee's circular, but would soon place America foremost in science among the nations of the earth.

Naval Hygiene, by Samuel A. Cartwright, M. D., of New Orleans, late of Natchez.

I. "The space allotted to each passenger." The existing laws seem to be defective on this head, leaving a wide margin for different interpretations. They estimate the space allotted to each passenger in superficial feet instead of cubic feet. They require fourteen clear superficial feet of deck on the lower deck or platform, but do not prescribe the height. Two or more tiers of berths may be superimposed ever so closely, even to suffocation, without violating the letter of the law—provided the lower tier be six inches from the floor or deck. The British laws place the limitation at ten superficial feet of deck to each passenger. They prescribe the height at not less than five and a half feet, interdict the erection of more than two tiers of berths, and forbid any ship from carrying passengers at all which is less than five and a half feet between decks. Whereas, our laws not only admit ships less than five and a half feet to carry passengers, but do not interdict, in express terms, the erection of a second tier of berths. Passengers are allowed to be carried in the orlop deck by giving each thirty superficial feet. More depends on the situation of the chambers occupied by the passengers, and the facility of keeping them dry, and of admitting air and light, than upon the amount of space allotted to each, or the number of passengers to the ton. Hence it is impossible to define the exact number that any apartment of given dimensions may safely contain, without a special examination of the apartment itself, its situation, the cargo, and all the circumstances connected with it, which had better be left to the physician of the port or some other officer. Mere science prescribing rules drawn from general principles, without taking into consideration all the attending circumstances, might suffocate with heat, and smother with their own exhalations an hundred in a chamber fully as capacious as another differently situated, containing twice the number, who, so far from being suffocated from the want of air, might have too much of it in the shape of piercing cold winds blowing through their apartment. Although the exact space on ship-board, necessary for each passenger, cannot be determined *a priori*, yet the requisite dimensions of the berths and standees can be determined with much more accuracy. They are not large enough under the present laws. They are required to be not less than six feet long by eighteen inches wide. This does not give room enough to a common sized man, and cramps and confines a large one in too small a space. The berths should be sufficiently long to enable their occupants to stretch out their limbs at full length; otherwise the unfortunate passengers, not not being able to turn over or extend their limbs, would become the subjects of that disease, very common at sea, vulgarly called *false-fat*, which is an œdema or celular infiltration from torpor of the vinous and lymphatic circulation, caused by confining the body in too small a space, not admitting of a full and free extension of the limbs. The berths, and also the temporary ones, called standees, should, therefore, be not less than six feet three inches long and twenty-one inches wide. Six and a half by two would be still better. A few of the latter size, at least, should be required for the accommodation of large men, while those intended for mother and child should not be less than three feet wide.

II. "The quantity and quality of the provisions allowed for each passenger."

As to quantity, the allowance under the present laws, both of the United States and Great Britain, is deficient. A healthy man requires a pound and a half of farinaceous or amylaceous food, and half a pound of animal food per day, or 14 pounds per week. Our laws require only ten pounds for each passenger per week; of which only one pound consists of animal food. The British omit animal food entirely in the allowance to emigrants, but require 11½ pounds of vegetable diet to each per week, besides two ounces of tea, viz: bread 2½ lbs., rice 2 lbs., oat-meal 5 lbs., flour and sugar, of each 1 lb. The American allowance consists of bread 2½ lbs., potatoes 3½ lbs., oat-meal, flour, peas or beans, and salted pork, of each 1 lb.; and sugar and molasses, of each ½ lb. The weekly rations of soldiers seldom contain less than 14 pounds per week, of which 10½ pounds generally consist of farinaceous and amylaceous substances, and 3½ pounds of animal food. The minimum rations on our southern plantations, for each individual per week, weighs 15½ pounds, of which 3½ pounds consist of salted pork, and 12 pounds of Indian corn meal, besides an indefinite quantity of vegetables, as potatoes, beans, peas, pumpkins, turnips, and many other things, including fruits, which are cultivated in their own or the kitchen garden, and are not estimated in the allowance at all. Indeed few of them are satisfied with 3½ pounds of salted mess pork per week, and many get 5 pounds. Napoleon was well aware that health, energy and courage depended on a good diet. The soldiers under him were allowed 250 grammes (about half a pound) of meat per day and one and a half pounds of bread. What is called half diet in a well-regulated hospital, is about equal in quantity to that required to be furnished to emigrant passengers by the laws of the United States and Great Britain, and which the emigration companies seldom exceed in laying in their ship stores. Unfortunately for the emigrants, their diet is not only deficient in quantity but is sadly deficient in quality. Provisions deteriorate or spoil on ship board much sooner than on shore. The inferior qualities are the soonest damaged. Metals also oxidize with more speed on ships than on land. This is not owing to the greater humidity of the open air at sea, which is about the same as the air on land, but to the greater humidity of the air contained in the ship itself. The air of a ship is much more moist, and somewhat hotter, than the external atmosphere. Some chambers are hotter and more moist than others. The one called the hold, for instance, is more humid by ten degrees than the outward air, and from three to four degrees warmer. From not paying attention to these facts, confirmed by the scientific observation of the surgeon of the frigate Herminie, in 1838 and 1839, having an equipage of upwards of 500 men, belonging to the French expedition against Vera Cruz, the stores of provisions soon became corrupted and deteriorated in quality from being deposited in improper places, exposed to a hot, damp, stagnant air. Oat-meal, flour, bread, potatoes, &c., soon become musty or mouldy if deposited in the hold or other close places between decks.

Provisions are not only soon deteriorated in quality by heat and moisture, but also if exposed to human emanations in a concentrated form, as they always are when deposited in the steerage of a crowded emigrant ship. No one can eat with the same relish in the wards of a hospital as in the pure air. The food partaken of in such places is so deteriorated in quality, by its coming in contact with the effete matter contained in the air thrown off by the sick, that it is digested with great difficulty, and is apt to cause vomiting or diarrhœa; hence the intelligent physician always hastens his patient out of the wards of the hospital or sick-room when the appetite begins to return. Seguin has found by experiment, that 18 grs. per minute are thrown off from the body in the form of what is called insensible perspiration, 7 grs. by the lungs and 11 grs. by the skin. The insensible perspiration consists of organized matter, already so altered in its properties as to be unfit for further purposes in the animal economy. When confined in a close, damp, hot atmosphere, it speedily corrupts and becomes fungoid; fungoid matter consists of living organisms. The practice of depositing provisions in such an atmosphere, or even bringing them into it after they are cooked, greatly tends to deteriorate their quality; it is like mixing good flour with that from the damaged fungoid grain called sick wheat. That scurvy, cholera, dysentery, ship fever, and other pestilences are mostly caused by eating food deteriorated in quality, drinking bad water, and breathing an atmosphere loaded with the moisture of human exhalations, there can be no doubt. When it is considered that every adult individual throws into the air around him about two pounds of effete matter per day, the importance of free ventilation will be perceived, as it is impossible to breathe or to eat in such an atmosphere without admitting the re-entrance of this cast-off matter into the human body again through the lungs and the stomach.

The articles furnished as provisions for the passengers and crews of ships should be selected from those which are wholesome and nutritious, and, at the same time, the least liable to be damaged by moisture. Bread made from flour, three parts wheat and one part rye, does not spoil as soon as that made from the flour of wheat alone. The preparation of cabbages, called saur-kraut, is an excellent anti-scorbutic, and would form a good addition to the allowance of salted beef or pork. Pickled beets, a valuable and cheap article, also sugar, tea, and coffee, should constitute a part of the provision stores. They keep better than most other things. Mustard, pepper, and horse-raddish root, are excellent condiments. An anti-scorbutic wine and syrup are prepared from the horse-raddish—the *cochlearia aromatica*. The allowance of sugar is too small; besides, few ships are ever supplied with the right kind of sugar. The maple-sugar of the north and the cane-sugar of Louisiana are greatly to be preferred to the West India sugar. As to maple-sugar, its healthfulness is inferred from observation of its effects. But in regard to the Louisiana sugar, its superiority to the cane-sugar of tropical climates is not only infered from observation of its effects, but is demonstrated by the new science of occular chemistry, as, also, by the usual method of chemical tests. In a hot climate, as that of the West Indies, the cane-juice speedily begins to decompose, and before it can be manufactured into sugar, an unhealthy product of fermentation is formed, known as uncrystallizable sugar, which, mixing with the good, gives the whole mass a disagreeable nauseating odor. The sugar made in a tropical climate always contains a portion of the above-mentioned unhealthy substance, called *lævogyrate* sugar, from its rotating the plain of polarization of polarized light to the left, when subjected to the test of polarized light. Whereas good sugar rotates to the right and is hence called *dextrogyrate*. In the cool frosty weather of November and December, when the sugar of Louisiana is made, the weather being cold, like the March weather when the maple-sugar is made, prevents fermentation, and consequently prevents the formation of that unhealthy product above mentioned, universally present in all raw sugars of tropical climates—which even the process of double refining cannot entirely eradicate—as proved, by the best double refined loaf being found to contain it. The process of refining destroys the aroma on which the virtues of cane-sugar, as an antidote for the narcotic properties of tea and coffee, depend. Being not only better, but cheaper than loaf or the clayed sugar of Havana. The American articles should be selected in preference to all other kinds for use on ship board. It crystallizes in coarse hard grains, and feels, when rubbed in the hand, very much like coarse salt, whereas impure sugar has a soft unctuous feel and a disagreeable odor. Good sugar is not only anti-scorbutic, but possesses within itself all the necessary elements for the nutrition of the human body. In M. Homberg's ship, sugar alone saved the crew after the provisions failed. (See Dictionaire Des Sceances Medicales, vol. 53, page 149.) Mixed with water it forms an agreeable beverage. It does not spoil like oat-meal, potatoes, and such things, but the longer it is kept the better it gets. In this it resembles good pork put up with solar salt immediately after the hogs are killed and the blood well drained off.

If the weather be warm, or the salt contains those impurities called *slack* and *bittern* by the salt-boilers, the pork, in a long voyage in a hot climate, will corrupt and produce the scurvy. Irish pork, which is always packed in solar salt, is improved by a voyage to the East Indies and back, whereas pork packed with Liverpool salt, or any other, however pretty and white in appearance, that is made by boiling, is sure to become sour and unhealthy at sea, as it is also very apt to do in the summer season on southern plantations, causing much sickness among the negroes. This is owing to the slack or bittern it contains, substances as unhealthy in salt as lævogyrate is in dextrogyrate sugar. There is no sufficient reason why ships should not be supplied with preserved meats and desiccated vegetables put up in boxes, canisters, and glass jars, hermetrically sealed. Their high price has heretofore prevented them from coming into more general use.

The desiccated cabbages, carrots, and green peas of Masson, and the potatoes and turnips of Gannel, were in 1852 reported on favorably by two boards of navy officers, appointed to examine into such alimentary substances, and of their adaptability to the navy. The price of some of them is very little beyond that of crude articles. The preparation of such substances on a large scale, for the supply of ships, is demanded by the progress of the age. Mr. Boland has an extensive establishment in Texas for preparing preserved meat out of good fresh beef. The practicability of preserving fresh meats for an indefinite period is no longer a matter of doubt. For more than a century goose-liver pie, cooked in Paris, has been a favorite dish in New Orleans. If the members of the boards appointed in 1851 and 1852 by the Secretary of the Navy, the Hon. William A. Graham, to inquire into the "important discovery" (set forth in S. Doc. 1, 1852, p. 485, of the President's message,) in regard to the questions of preserving peas, beans, sprouts, carrots, turnips, &c., in a fresh state for months and years, and the practicability of embalming fresh meats, had convened in New Orleans and called for dinner at a French restaurat, they would have had all those questions satisfactorily settled by the time they had finished their meal. Such articles are daily served at all the principal restaurats in the French part of the city, at a price not exceeding the usual hotel rates for the same species of substances just from the vegetable market and the butcher's stall. In regard to the question, whether the United States could afford to purchase such articles for common use as diet in the navy, the boards of naval officers above mentioned could easily have satisfied themselves by visiting almost any French settlement in Louisiana, where they would have found them on the tables of many of those who consider themselves too poor to send their children to school, and believe that a knowledge of the English language is an accomplishment like that of Italian, which none but the wealthy can afford to incur the expense of acquiring. By examining the gardens and surrounding fields in the French settlements, they would have discovered many of the articles which the inhabitants prefer importing from France, already prepared for table use, than going into the hot sun gathering and preparing them with their own hands. The vegetables are cheap, but the preserved meats are dear. It would be cheaper, however, to purchase a small supply of them for use at sea than to depend upon coops of live poultry. Poultry, pigs, &c., when taken to sea, become sickly, and are not good to eat. The same thing happens to boat chickens, as they are called, brought to New Orleans from the States on the Ohio river. The bread baked in London and sold in this country under the name of Cracknell biscuit, at the high price of $1 25 per box of three pounds and three-quarters each, could, no doubt, be manufactured on a large scale in the northwestern wheat-growing States at a price so reduced as to admit of its general use at sea, and would form a most excellent substitute for other bread, which is so apt to spoil or become mouldy.

The quantity of water required by law to be distributed to each passenger, per day, is only three quarts. A gallon and a half would not be too much. What is worse is its general impurity. It may be laid down as a very useful, practical, general rule, that any water whatever, which will not mix readily with soap, however clear and free from any unpleasant taste, is unfit for use on ship-board; also any which has an unpleasant taste, however well it may mix with soap. As ships cannot always be supplied with good water, they should be supplied with a little pure sand and charcoal, for the purpose of having all hard or bad-tasted water filtered through these substances previously to its being used by the passengers and crew. It should also be cooled with ice. Cold causes the unhealthy organic matter contained in water to be deposited.

The supply of vinegar—a pint a week—required by law for each emigrant, is more than necessary. Too much vinegar is unhealthy. A gill would be sufficient. In place of the other three gills, a quarter of an ounce of citric acid, in chrystals, could be allowed. The citric acid, or concentrated lemon-juice, mixed with sugar and water, in the proportion of twenty grains to a pint, and flavored by a bit of lemon peel or a drop of the oil of lemon, makes a very pleasant and healthy lemonade. Half an ounce in a gill and three quarters of water, makes a solution of the average strength of fresh lemon-juice. With a few grains of salt-petre and a little vinegar, the citric acid forms a most efficacious and agreeable remedy for scurvy. It is also used in forming the neutral mixture and effervescing draft in the treatment of fever. It has no other advantages over lemon juice than uniformity of strength and not being liable, like the juice, to spoil.

III. "The permission allowed to the passengers to furnish their own provisions for the voyage, instead of making it, in all cases, the duty of the master to provide them."

The furnishing of proper supplies of provisions for a sea voyage, whether for one individual or a thousand, requires a great deal of knowledge and experience of a peculiar kind, unobtainable anywhere else than at sea and in the great seaports. Even in those places it requires a well-disciplined mind to gather it. The bare statement of this truth is sufficient to show that the emigrants, who are mostly illiterate people, and know little or nothing about ships or nautical affairs, are not competent to furnish themselves with the necessary provisions for the voyage. What is necessary for all is necessary for each. So far from passengers being permitted to furnish themselves with their own provisions, it is very doubtful whether the officers on ship-board called storekeepers, (who generally make the purchases,) or the master, or the agents of the shipping companies, should be permitted to furnish the supplies for their own ships. But the American and British governments distrust their fitness so far as to prescribe rules to govern and direct them to a certain extent. Thus, they are required to take on board certain specified quantities of provisions, of many varieties, expressly named. But the question is, whether it would not be better to deprive them of the permission to purchase the required provisions where they please, governed only by their own judgment of their quality? It would be probably better to give the whole business of supplying ships in foreign ports to the consul, and at home to agents appointed by the government. Each ship would then be supplied with all necessary articles in the provision line, approaching much nearer to the same quality than they now do. At present, the quality depends upon the ship's storekeeper, master, or agent, and the merchant furnishing the supplies. It must, therefore, greatly vary. But if the consul, or some trustworthy agent of the government, had to select the supplies, there would not only be more uniformity, but the quality would be better and the price less. Large purchases can generally be made on better terms than small ones. The agent or consul, residing constantly in the same place, would have a better opportunity of knowing where and of whom the best quality of provisions could be obtained than a comparative stranger. The master or agent, however, of every

9

ship should have the privilege of purchasing where he pleased any additional provisions of any kind he might see proper. So, also, the passengers should have the same privilege of supplying themselves with anything extra they might choose—intoxicating drinks alone excepted. Much unnecessary expense is incurred from the spoiling of provisions on ship-board. This often arises from that measure of prudence, requiring more to be taken on board than will be probably needed during the voyage, in order to have a sufficient supply in case of accidental detention from calms or other causes. The surplus quantity, having made one voyage, is generally damaged before it makes two—the damaged articles being a total loss. Such losses could be avoided by taking on board no surplus of the usual articles entering into the allowance, but to guard against accidents by a supply of preserved meats and dessicated vegetables, to be used only in cases of scarcity and necessity. Such provisions, not being liable to spoil, would always be ready for use in subsequent voyages. In the navy, a large amount of provisions, in the public storehouses and on ship-board, is annually spoiled and thrown into the sea, or condemned in consequence of having been kept too long in reserve to meet any unforseen scarcity. Medicines are also apt to spoil. Fortunately, there is one medicine, the best one, too, for the usual complaints at sea, which never spoils or is found wanting. That medicine is sea water. It is both emetic, aperient, and tonic, according to the dose. Sea-sickness cured by it rarely troubles the patient again. It is the best remedy for the constipation, so common at sea, caused by the rocking of the vessel, diminishing or suspending peristaltic motion of the bowels. The rocking chair, so much more in use in this country than in any other, has the same effect, (hence standees have very properly been substituted for hammocks.) More than a hundred years ago, Dr. Richard Russel highly recommended sea water as an internal remedy for a multitude of diseases, in a work entitled "*De usu aquæ marinæ.*" Two thousand three hundred years ago its virtues were extolled in Grecian verse :

Mare oblujt omnia hominum mala.

Euripides' Iphigen. in Taur., *V.* 1193.

CANAL STREET, NEW ORLEANS, *February* 27, 1854.

Naval Hygiene, by Samuel A. Cartwright, M. D., New Orleans.

IV. "Ventilation." A house, or booby-hatch, over each passageway leading to the apartments allotted to the passengers, with doors and windows in it, to be left open for ventilation, is required by law in the construction of all emigrant vessels. Also, two ventilators in every apartment containing one hundred persons. In addition to the usual ventilators, a ventilating stove, on the principle of that of Wuettig, would be very useful for ventilating and drying the atmosphere of the hold and other dark, damp chambers. In order that the animal and other exhalations should be conducted off as fast as formed, the newly discovered ventilators, having tops closed above and open at the sides, made of iron and passing through the decks down in the apartment to be ventilated, promise to be of great service.

The ventilators called wind sails, made of canvass, fastened around strong hooks, to keep them from collapsing, and having a broad mouth, kept open by ropes attached to the rigging, with the other extremity opening into the chamber below, have so greatly contributed to preserve health on board of the ships which have used them, that they are now adopted in the United States navy.

V. "The Cooking Arrangements." The camboose, or cooking range, should be on deck and under cover. None but experts should be permitted to cook. There could be no objection to one or more passengers assisting and acting under the direction of the regular cook, if their services were needed. In the Dutch ships, the camboose is between decks near the centre of gravity. Rough weather does not so much interrupt the culinary processes when carried on in the central part of the vessel. The cooking fire, also, serves to expel dampness, but owing to the danger of accidents by fire and to the annoyance of the smoke, the cooking had better be performed elsewhere.

VI. "The duty of the master to enforce personal cleanliness, and to insure the cleanliness of the vessel." It would be utterly impracticable to enforce personal cleanliness without conferring on the master a power equivalent to that possessed by a colonel over his regiment, unless some plain and specific laws be enacted to embrace the subject. The decks of a ship may be ever so clean and yet the vessel foul, if its contents, including the passengers, their clothing and bed clothing, be not also clean. The masters of vessels report, that so great is the aversion of many emigrants to wash themselves or their clothes, or even to come on deck a sufficient time for the steerage to be cleansed, that actual force has often to be used with ropes, or other means, to hoist them on deck. Not unfrequently they meet force by force, believing that the master has no authority to compel them to leave their room, or to get out of their berths. Hence fights and disturbances often ensue between the passengers on the one hand, and the crew on the other, in executing the orders of the master to cleanse that part of the ship occupied by the emigrants. The accumulation of a large number of individuals, for days and weeks together, in a small space, is known to corrupt the air and to generate an artificial malignity. Unless the master has the power, by voluntary agreement or by law, expressly conferred on him, to disperse the crowd thus confined in a small space, to make them exercise and air themselves, he cannot prevent the generation of an artificial and an unhealthy atmosphere in their chambers. Transport and other troops, when kept in motion, are always more healthy than when permitted to remain too long in their hammocks or standees. The greater the number and the smaller the space the greater is the necessity for exercise. If they only have room to turn round, or to march in lock-step, it is better than to be still. Men, like water, generate corruption when still, but dissipate it, as fast as generated, when in motion. Pestilence does not afflict soldiers on the march, but only in camps or barracks. Emigrants would be much healthier if they were compelled to come on deck and promenade in regular order, as soldiers on the march—holding one another up if the sea be rough. The sea-sick should be handed over, if there be no surgeon on board, to the care of some old, discreet and faithful sailor—

the best substitute for a surgeon on such an occasion; the patient, after going through a course of sea water and afterwards promenading the deck in the fresh air, supported by one or two of the crew, or some friendly arm, to keep him from falling, will soon be well. So great is the disinclination of young sailors and mariners to drink sea water, and after it operates to stir about in the open air, they would be apt, like many of the passengers, to be afflicted with sea-sickness during the whole voyage, and perhaps subsequent ones, unless they were compelled to take the medicine, and after a short time to leave their hammocks and exercise in the open air. After sea-sickness has abated, there still remains so much disinclination to motion as to require the stimulus of compulsion to induce the patient to continue the exercise, until the *mal de mer* leave him entirely, or it will return with more or less violence. Exercise in the open air, besides being a preventative and curative remedy for sea-sickness, is healthy and beneficial for all, in good weather or bad. The top-gallant men, exposed to all kinds of weather, are the most healthy portion of the crew; their duties often requiring nearly every muscle of their bodies to be called into action. Clinging to or balancing on a rope, gives more exercise than on a wide and secure surface, as it calls into play a greater number of muscles to maintain the position. So, also, walking erect and in regular order, gives more than twice the exercise of the same amount of walking in a crooked, slovenly manner. In the latter case only a few muscles—those absolutely necessary for progression—are called into action, while in the former more than twice the number are thrown into tension to hold the head and trunk erect, expand the chest, and give precision and regularity to the movements. The power, therefore, to enforce exercise of any description, as that of walking, would be apt to fall short of the object for which it was conferred, unless it could enforce *the manner* in which the prescribed exercise was to be taken. Walking in a doubled up, slovenly manner, without expanding the chest, instead of being a healthful exercise, is almost the same thing as no exercise at all, because it brings so few muscles into action.

A regular inspection of the emigrants twice a week, at least, would greatly tend to promote health and cleanliness. But it would be worth nothing unless there was some express power conferred on the master to have those punished, in some way, who did not bear inspection. This power might be conferred directly by law, or indirectly by the emigrants voluntarily consenting to concede to the master the power, by signing an agreement to that effect, previously to coming on board. The scarcity of fresh water, and the difficulty of drying clothing washed in sea water, are unfavorable to cleanliness at sea. The chloride of calcium contained in sea water is an extremely deliquescent salt, and attracts moisture with so much avidity that there is much difficulty in drying clothing washed in water containing it. What is worse, the clothing, after being dried, becomes damp again on being worn a little while. This inconvenience can, in some degree, be overcome by adding to the sea water a little soda, which decomposes the chloride and disposes the articles washed in it to dry with more facility, besides preventing them from becoming damp again after being put on. The soda also favors the dissolution of the soap. Every ship ought to be provided with a sufficient supply. There is nothing better in naval hygiene to dissipate ennui, and to dispel the sad affections incident to a monotonous sea voyage, than gayety. Gayety is the natural remedy for sadness. The emigrants should be encouraged to enjoy themselves on deck with music and dancing—under a power, however, to enforce order and decorum as strict as that over soldiers on parade; otherwise such amusements would become a nuisance.

New Orleans, *February* 27, 1854.

Naval Hygiene, by Samuel A. Cartwright, M. D., New Orleans.

VII. "The employment of a qualified and experienced surgeon."—This would greatly add to the expense, and be of no avail to the larger portion of the emigrants, unless the ships were better constructed to accommodate that class of passengers, and the whole were under something similar to military regulations, and could be made to take medicine and follow advice. The writer was once a passenger on board a ship having a well qualified and an experienced surgeon, and a great number of emigrants closely packed in narrow berths, in three rows, one above another, in a dark, damp, ill-ventilated apartment, abounding in filth of the most disgusting kind. At length an epidemic broke out among them. The efforts of the surgeon to save them were most praiseworthy, but were rendered abortive by the noxious vapors that his patients were compelled to breath, and by their own ignorance and obstinacy in refusing to take medicine or listen to advice until it was too late. Ever and anon some one was plunged into the sea, sewed up in a hammock with stones at his feet. If near the moment of dissolution the unfortunate emigrant had consented to take anything in the shape of medicine, the medicine was unjustly blamed by the survivors with causing the death. The writer being called on for advice, urgently recommended the removal of the sick and the well upon the spar-deck into the open air. The measure, as soon as carried into effect, checked the epidemic and no more deaths occurred. The prevention of disease and the preservation of life at sea depend less upon the medicine chest than upon fresh air, cleanliness, and good accommodations. The atmosphere of the ocean is purer than that of the land. If the ship, crew, and passengers be clean, the provisions of sound quality and the water good, there will be nothing wanting to preserve health but plenty of exercise, free ventilation, dry rubbing of every apartment of the vessel, brushing and airing the clothing, together with gayety and cheerfulness—provided the ship be not overcrowded, and no contagious disease be admitted on board, or any person from the infected atmosphere of a foul jail or hospital. So often has observation declared that a ship, which takes on board a passenger immediately from a jail, hospital, or other sickly, crowded place, is apt to be infected with a mortal malady, after being two or three weeks at sea, that more good could be done by a stringent law preventing such passengers, or their clothing and bed clothing, from being taken on board until after a thorough washing and ventilation, than by requiring the presence of a qualified and experienced surgeon. The object in view of that learned body, the American Medical Association, in appointing a committee to memorialize Congress in favor of a law requiring every emigrant ship to have a surgeon on board, was most praiseworthy. It was dictated by a knowledge of the great mortality among emigrants from ship fever and the intractable nature of that malady. It was thought that if well-read physicians, under the name of surgeons, were to be always present on emigrant ships to watch the rise and progress of that pestilence called ship fever, that some one or other of them might hit upon a more successful practice in the treatment of that terrible disease. The association well knew the risk that physicians would run in making their bed in the nest of pestilence, but were nobly willing to sacrifice a portion of their own numbers for the public good.

One important fact, however, was overlooked—that ship fever is an artificial pestilence, scorning physic, and that the United States Congress is the best doctor for it. Already has our system of government, guided by the light of medical science, smote the jail fever, and banished it from America. It has also made such a thorough sweep of that immense class of disgusting, artificial ailments, usually denominated surgical, that no professed surgeon, throughout the length and breadth of the United States and territories, except in a few crowded, ill-governed cities, can gain a livelihood by the practice of surgery proper, if he depends upon his own neighborhood alone to supply him with patients. Yet, in ill-governed Europe, every neighborhood—even those unable to support a physician—has one or more surgeons, more or less overrun with practice. Accidental injuries form but a small portion of surgical diseases. America has her due proportion of wounds and bruises from accidents; but her institutions have been the balm of Gilead for those putrifying sores, and other offensive, crippling ailments, denominated surgical. Equal laws and light taxes have dried up the fountain of surgical maladies, and left the surgeon proper without a vocation on this side of the Atlantic. The fountain that fed them was an impoverished state of the blood, caused by a deficiency of food, fuel, clothing, fresh air, and the comforts of life. Ship fever can be banished from the ocean by the same means that jail fever, and the whole host of surgical diseases, were cast out of the United States of America. No art was a match for the pestilence generated in the dark, dank, crowded cells of filthy, unventilated jails—the inmates eating unwholesome food, drinking impure water, and re-breathing the foul emanations of their own bodies. No art can cope with a similar pestilence, generated from similar causes, on ship-board. Medicine is a branch of the natural sciences, operating by natural means. Artificial diseases, as ship fever, and those pestilences which sweep off the poor and destitute in crowded, ill-ventilated cities, are not obedient to natural means. The true method is to banish them, by destroying their artificial causes. As soon as a generous public diffused the comforts of life among the seventy thousand destitute emigrant population of New Orleans, last summer, the pestilence, which was sweeping into eternity three hundred a day, immediately began to disappear, before frost, or any change in the weather—its artificial fabric being broken down by the beneficent hand of the American people.

In striking down the artificial causes of ship fever, the physicians of the Medical Association can render important aid to the ship-builder, the masters and owners of vessels, and to the governing authorities, and thereby greatly promote the interests of humanity. They can demonstrate that man requires a certain amount of oxygen, and if he does not get it, he sickens and dies in the presence of the ablest physicians of the land; that there is no medicine of sufficient virtue to enable mankind to do without wholesome food, drink, and fresh air.

The practice of building ships, even those of our navy, so constructed that the higher officers and cabin passengers have state rooms and saloons unnecessarily capacious,—ostentatiously so,—at the expense of the space that rightfully belongs to the crew and to the emigrant passengers, is at war with the first principles of hygiene, as it evidently is with the first principles of republicanism. It is, moreover, muzzling the ox that treadeth out the corn. So far from this practice, in the construction of sailing and steam ships, going out of use, it seems to be daily coming more and more in vogue. Unless arrested in the navy, that arm of our national defence will soon be deprived of American seamen altogether, as none will be found willing to be housed worse than dogs in a kennel, and not fed so well, that the officers and favorite passengers may live in palaces and fare sumptuously.

The practice of crowding seamen into narrow, dark, damp cells, excluded from the fresh air, is already causing loud cries to come up to Congress for more marine hospitals in every direction. Hospitals are nothing but anti-republican sores on the body politic. The true policy of our government is to remove the necessity for such institutions. All Louisiana, including every city, town, village, and parish in the State, does not send more than one hundred and fifty patients of her native people to the great Charity Hospital of New Orleans per annum; whereas the various wretched governments of Europe crowd its wards with from twelve to fifteen thousand patients annually. Thus proving that our happy country has little or no use for such pestiferous establishments, which despotism and oppression have made the only homes of countless multitudes of the laboring classes; where a large proportion of them are born, and about a third of them die.

It is in the mission of every representative sent up to the State capitols and to the federal government so to legislate as to avoid doing anything which has a tendency to drive any portion of their constituents into the hard necessity of seeking such homes.

The medical profession will not assent to any measures creating the necessity for hospitals, by depriving seamen on ship-board of the space necessary for healthy existence, to make room for gaudy saloons. Nor ought it to consent to be made the scape-goat for the sins of avaricious ship-masters and emigration companies, by sending surgeons on board their over-crowded ships, to bear all the blame of that excessive mortality produced by want of proper food, good water, and fresh air, which no skill in medicine and surgery can prevent, and no power but that of Congress can remove.

VIII. "The employment of a reasonable number of attendants to minister to the sick, and to enforce the observance of cleanliness, both of the persons of the passengers and of the vessel."—It is believed that persons specially employed for such purposes would be unnecessary, and would add greatly to the cost of passage; so much so as to prevent a majority of those wishing to emigrate, from ever reaching the asylum of the oppressed. Congress, once inveigled into a course of legislation having the effect of arresting emigration by increasing the cost of passage, would soon be besieged with petitions to pay the additional expense incurred by the employment of surgeons, nurses, and supernumerary attendants, required by the laws. This would add to the profits, without correcting the evils, caused by the greedy avarice of certain unprincipled men engaged in the business of transporting emigrants. It would be better for the United States to increase the navy, and to give the emigrants a free passage in our vessels of war—selecting husband, wife, and children, and single persons of good character; rejecting the vicious tenants of prisons, and the sickly, disabled paupers of poor-houses and hospitals. There is no necessity of employing regular nurses on ship-board. The object in view would be better accomplished, and with less expense, by conferring authority on the master of every vessel, to select nurses for the sick from among the emigrants themselves. When thus pressed into service, they might be allowed a fair compensation, or each, in his turn, to act as nurse without pay. Unless the proposed regular nurses were polyglots, or conversant with the languages and various dialects spoken by the emigrants, they would be unsuitable attendants for the sick—not being able, even, to understand that dialect of the English spoken by the emigrants from Cornwall. Nor could menials or petty officers enforce the observance of cleanliness, which is more than the master of the vessel can do, unless aided by some plain and express law of the United States, giving him the requisite authority.

—

Naval Hygiene, by Samuel A. Cartwright, M. D., New Orleans.

XI. "A report to be made by every vessel bringing emigrant passengers of the length of the voyage, number of passengers, number of deaths, &c., to be published, and to be returned to the State Department."—Such reports, to be useful for statistical purposes, should be accompanied with provisions to insure accuracy in their details. The British have a regulation of the kind, but it is so loosely carried into effect, that the ages, and even the sex, are not always given.

XII. "In case deaths have occurred during the voyage, an inquest to be held under the supervision of federal officers, and the verdict to be published, and returned as above."—The utility of such an inquest is not perceived, unless in cases of doubt or suspicion.

XII. "A limitation of the number of passengers allowed in every vessel, in proportion to the tonnage of the vessel."—The present limitation, two to every five tons; the British regulation is three to every five tons, including the crew. Our laws admit fewer to the ton than the British, when interpreted by tonnage; but there is more difference between the two countries than appears on the face of the laws. The former computes the tonnage by taking half the breadth of the ship for the depth, which, in deep vessels, falls greatly short of the actual tonnage—while the latter estimates it by guaging the ship as a cask is measured. The American tonnage, in deep vessels, often exceeds the actual, or British tonnage, fifty per cent. or more. In very shallow vessels, however, it is even less than the British. But that no safe rule can be deduced from the mere tonnage of a vessel is proved by the fact, that ships, which bring to the port of New Orleans three or four hundred emigrants each, have not room, on their return trip, for a single passenger; the master himself often giving up his apartment, and sleeping on a bale of cotton, to make the more room for the freight, which fills the whole ship. The standees, or temporary berths, are removed, and the steerage, cabin, and every vacant place, is packed as full as it can hold with cotton bales.

CANAL STREET, NEW ORLEANS, *February* 27, 1854.

Naval Hygiene, by Samuel A. Cartwright, M. D., New Orleans

IX. "The separation of the sexes, and the prevention of unnecessary intercourse between the crew and the passengers."—It would be inexpedient, if not impracticable, to separate the sexes by dividing families. The mother with a number of children requires the presence of the father to assist in taking care of them. Or if either parent were permitted to take the children in his or her apartment, the question would arise in regard to the age at which mankind cease to be children and become adults. As the time varies greatly, according to the sex and constitution, much difficulty would arise in applying any general rules to the subject The difficulty could be overcome by allotting the central portion of the apartment to families, and one extremity to unmarried females, and the other to the single men. Any improper intercourse between the crew and the passengers, or the passengers themselves, should, of course, be prevented as far as order, system, and authority, can prevent evil. But there seems to be no sufficient reason why the crew should be prohibited, by law, from partaking in the amusements of the passengers, when off duty, if invited by them to do so, or of holding conversation with them, on the condition that order and decorum be strictly observed. The room on ship board for exercise is sufficiently contracted without making it more so by restricting different parties to special localities in the unoccupied space. It is not in conformity with the spirit of our institutions to make differences in degrees when nature and education have made none. The rude and uneducated would not relish the society of the refined and polished, nor would the latter seek the companionship of the former. But there is no good reason why legislative power should be invoked to cut up each of those two great natural divisions of society into artificial orders, making distinctions where none properly exists. It is true that sailors have a specific character, but this is a consequence of their mode of life. Their seeming rudeness is caused by their being habituated to live apart from the company and conversation of the gentler and more amiable sex. Like other men, they would exchange rudeness for gentleness and delicacy in the presence of virtuous women, rather than be punished by being deprived of the pleasure of their company. It is the women, and not Congress, who should give the law on this subject. The idea that seamen would be more unmanageable if permitted to enjoy any privileges not inconsistent with their duties, is not of American origin. Though rude, American sailors, at least, are not ferocious, but humane, chivalric, and generous. Though little sensible to their own evils, they are not so to others. In ship-wrecks, and other casualties at sea, they perish themselves sooner than let a woman drown, or relax an effort to save the ship and passengers. They make it a point of honor to be the last to leave a ship in flames or going to pieces on the rocks. They are never disobedient to orders when the tempest howls and danger threatens. It is not from fear of punishment that they are true to their profession under such appalling circumstances, but from a deeper principal radicated in that moral virtue which none know so well as the American mother how to instil into the human heart. The great medical author, Mosely, in his work on Tropical Climates, (third edition, page 49,) gives an instance of the display of that principle in an American seaman captured during the revolutionary war by the British frigate Pallas, and ordered to serve the enemy against his own flag. No torture or fear of punishment could induce him to obey such orders. He jumped overboard, a shark seized him, and tore the flesh from both legs, taking off a foot at the ankle. When dying he was rescued from the monster, and told the good Dr. Moseley, who hastened to his assistance, that he "was happy in laying down a life he could not employ against his country's enemies." Any legislative action imposing unnecessary restrictions upon such men, as cutting them off, as proposed, from any intercourse the passengers might be willing or desirous to hold with them, would be inexpedient. It was Moseley's intercourse with such men that made him exclaim: (p. 154,) "Every man in North America is by nature a general." It was intercourse with such men that made him (although an enemy to republicanism, but a deeper enemy to despotic cruelty) see, confess, and record a great truth, in the year 1780, which is not even yet clearly seen and acknowledged by all of our representatives in Congress, viz: "*That extinguishing European tyranny in the western hemisphere is a debt which North America owes to the world.*"—(Moseley on Tropical Climates, page 153.) Intercourse with such men would lead to the prompt payment of that portion of the debt that North America owes to the world, which has been past due ever since the Spanish authorities in Cuba shot down like dogs, without a trial, fifty-one gallant Americans, for no other crime than presuming to imitate the example of Lafayette, among whom were no less than five worthy and noble members of the medical profession.

CANAL STREET, NEW ORLEANS, *February* 27, 1855.

Naval Hygiene, by Samuel A. Cartwright, M. D., New Orleans.

X. "A thorough process of disinfecting every vessel on board of which disease has once made its appearance."—There is no better process of disinfecting a vessel than that of soap and water, dry rubbing, and free ventilation. Any other is a delusion.

Three orders of substances infect the air of a ship, either dissolved or mixed in it: first, gasses which can be demonstrated by the eudiometer and other means, as the carbonic acid gas, for instance.

Secondly, substances which are inappreciable by instruments or chemical agents, but are cognizable to the senses, as various odors.

Thirdly, such that neither the senses nor chemical tests or instruments can detect, and are known only by their effects.

Disinfecting agents can destroy the carbonic acid and other appreciable gasses contained in the atmosphere of a room, and may also mitigate or remove unpleasant odors, but they cannot be relied upon to disinfect a vessel from those more frequent causes of disease which neither the senses nor chemical agency can detect in the atmosphere. Being unknown, the substances having the power to neutralize them are also unknown.

Nothing but a removal of the atmosphere by free ventilation, and a thorough scouring and dry rubbing of the unhealthy apartment, can be depended upon. Filling the air with foreign matters called disinfecting agents often does more harm than good, as it adds other impurities to the atmosphere.

Lime and its preparations, the chloride of lime, and many other substances, can destroy some noxious gasses, and arrest the process of putrefaction and fermentation, but it is much safer to remove the putrescent and fermenting materials from which the mephitic vapors and bad odors arise.

Even in the case of carbonic acid gas, the best known, and the easiest to be neutralized by chemical or disinfecting agents, ventilation, washing, and scrubbing are more effective than any other measures in destroying it.

It is very apt to form in the hold of a ship, being heavier than the atmosphere, it circulates with difficulty, and attaches itself to the sides of the vessel, and into angles and crevices.

Lime and ventilation and heating the apartments are all useful in attenuating it, but swabs wet in fresh water, followed by dry rubbing, are necessary to its total destruction.

Dry rubbing is not only one of the best means to remove noxious gasses, but is the best to remove their causes, and also to dry the atmosphere of an apartment.

Cook and Vancouver were very partial to dry rubbing. The usual practice of constantly inundating the decks and different apartments of a ship with sea water, and occasionally sprinkling them with chloride of lime, greatly tends to add to the humidity of the atmosphere within the vessel.

Neither of the celebrated navigators just mentioned was favorable to the common practice of frequent wetting and washing the decks with sea water, but preferred sand and dry rubbing. They put in practice the principles laid down by Dr. Rouppe. In his treatise on the diseases of seafaring men Rouppe says: "*Et quoties primum purgatur tabulatum, madidatur, quod tamen melius esset ope radularum purgare, sicco manete tabulato.*"

Truly it is better to keep the decks dry and cleanse them by dry rubbing than to wet them as often as they are cleansed.

Vancouver, tom. 1, pp. 30, 31, says, "Le soin de ne pas laver trop souvent l'interieur du vaisseau étaìènt précautions indispensables, et quil en résultait les effets le plus salutaires procu la santé de l'équipage."

But Stavorinus, of Bavaria, has met the question of the best means of disinfecting any vessel on board of which disease has made its appearance, and has practically solved it.

The moment any disease of an epidemic or contagious kind showed itself on ship-board, he had all the apartments of the ship cleansed and purified without employing water, or the so-called disinfecting agents, because experience in former voyages, he informs us, had proved to him that *humidity* was the chief agent by which the air becomes vitiated, and that its removal restores health. Humidity is the vehicle by which emanations from the body and foreign matters are diffused through the atmosphere.

It is likewise essential to putrefaction and fermentation. Its removal dries up the sources from which noxious vapors arise, by stopping the putrefactive and fermentative processes.

Apart from the unhealthy atoms to which it gives wings, humidity with cold induces great inaction of the vascular system, giving rise to scorbutic ailments, colds, influenzas, rheumatism, glandular engorgements, and adynamic fevers; united with heat it causes bilious, remittent, and yellow fever, dysentery, diarrhœa, colics, dropsies, &c. In both cases it relaxes the skin, suppresses the cutaneous and pulmonary exhalations, and increases the secretion of the mucous surfaces.

Hence, pulmony catarrhs, oedema of the lungs, dysnocæ rheums, and sore throats are so prevalent when the air is charged with too much moisture. The humidity of the atmosphere in ships, as a cause of disease, has not attracted sufficient attention in this republic, which has the largest mercantile navy in the world. In proof of which, it is only necessary to turn to the Navigation Laws, (Regulations of Passenger Vessels, 2589,) where it is made the duty of the captain to have the decks occupied by the passengers cleansed with the chloride of lime, a diliquescent salt in its impure state, attracting moisture and thereby breeding cholera and other ailments. When ever so pure, it soon becomes impure by exposure, and keeps the decks damp instead of drying them. Owing to the fact that excess of moisture in the atmosphere benefits some constitutions of a lean and arid fibre, its injurious effects upon the majority have, in a great measure, been overlooked; at least, so far, that no efficient measures have been adopted to prevent its excess in our shipping, but, on the contrary, customs and habits have been carelessly fallen into, and laws enacted which tend greatly to increase it.

The amount of humidity in the atmosphere is not a supposition, as it was in ancient times, but can be accurately weighed and measured. In early times it was one of the four great principles of the doctrine of man, to the excess or deficiency of any one or more of which, the ancients not only referred the origin of diseases, but also the different temperaments. Although wrong in theory respecting heat, cold, dryness, and moisture, they were practically right in their inferences touching the effects of their excess or deficiency.

Our countryman, Prof. Espey, has paid great attention to the subject of atmospheric humidity. The writer spent a whole summer, almost alone, with that distinguished meteorologist, from whom he could not fail to learn that humidity is closely connected with hygienic measures, particularly those relating to disinfecting houses on land and vessels at sea in which disease has made its appearance. That excess of moisture in the atmosphere has a powerful effect upon the human frame, rests not upon conjecture, but upon well observed facts.

Observations, however, until analyzed and arranged in connexion with others, have very little practical utility. Every one has observed the effects of moisture on curly-haired persons, and that in a humid atmosphere the strings of instruments, formed of animal tissues, stretch so much as often to break.

Delicate persons feel more languid and have less appetite and thirst in damp weather, nor is their digestion so good. The mucous surfaces become embarrassed with superabundant mucosities, causing invalids to haulk and spit more than when in a dry air. Their hearing, taste, and smell are also less acute; their motions are slower, and there is a want of that agility which a dry, bracing atmosphere gives. The passions partake of the general atomy of the body. Even children have observed that deaths mostly occur in stormy, rainy weather, without knowing that humidity, by relaxing the hold of the vital principle on the human organization, is the cause of that sad phenomenon being connected with a weeping sky.

The effects of humidity on the animal frame is also made apparent by anatomists using it to separate the fibres and tissues of the dead subject from one another in their demonstrations.

The causes of excessive humidity on ship-board are well worthy of investigation. The humidity varies in different apartments of the vessel. Péron, a surgeon on a French corvette, made repeated experiments with a hygrometer, and accurately measured the amount of moisture in different parts of the vessel. The hold was 12° more moist than the air of the surrounding ocean. Hence it was inferred that the leakage water, the want of ventilation, and the fact that sea water contains a portion of diliquescent salts, were the causes of the excessive humidity. Ventilation and keeping the leakage water as low as possible by pumping, removing the diliquescent salts by swabbing with fresh water, and afterwards by dry rubbing, are plainly pointed out as the best measures to guard against the excessive moisture in that part of the ship. He also found that the apartments in which the greatest number of persons were collected, were much more moist than those of the same size which contained a smaller number.

Human exhalations therefore may be set down as another source of moisture. He also verified the preceding observations, made by Dr. Rouppe, that want of ventilation increased both the temperature and the humidity of the air in which a number of men were confined.

Rouppe, off Maderia, had the hatches closed on 180 soldiers asleep in their hammocks; the heat of the atmosphere rose from 77° to 85° in a short time, and so much aqueous vapor accumulated that everything in the apartment was dripping with moisture. The sol diers' sleep became disturbed; they sighed deeply, and appeared greatly distressed.

He then let in the air; the moisture was soon diminished, and they slept calmly. This experiment was repeated several nights in succession with similar effects.

Péron found, by the hygrometer, that the dryest part of a ship-of-war is what the French call the *saint-barbe*, which we call the gun room. This room is never cleansed by washing, but only by dry rubbing.

As its humidity was very little greater than the external air, he inferred that dry rubbing greatly tends to remove moisture. It has also been observed that men in motion do not vitiate the air with moisture so much as when still.

These observations were made a long time ago, and have been repeatedly confirmed. Yet it does not appear that any other than a few celebrated and successful navigators have turned them to practical account in preventing disease and in disinfecting vessels.

Chemistry, soon afterwards, began to make great progress, and all eyes were turned to that science, hoping that some disinfecting agent would be discovered to neutralize or destroy the miasm supposed to be the causes of a large class of the most fatal diseases. One after another has sprung into notice—been tried and failed—scarcely leaving time to regard the true knowledge already gained as worthy of any attention in the race that all seem to be running after some newly-discovered disinfecting agent of the chemical laboratory.

Chemistry has made great progress, but it has not yet discovered any gas, except oxygen and its combinations, which will support animal life when respired. All cause death, if void of oxygen, as soon as breathed.

As oxygen is the only substance which is respirable at all, it would seem that it is the only one that gives any promise as a disinfecting agent. Atmospheric air contains about twenty per cent. of this gas, neither more nor less, whether on land or sea. Sickly places or healthy, the constitution of the air is chemically the same; but it is more absolutely pure, as it contains variable quantities of other matters and gasses. Many substances, however, abstract it from the atmosphere. Water has this effect. Oxygen has a very powerful attraction for most substances, and combining with them, oxydises them. An excess of oxygen, thrown into the air, might combine with those noxious matters which cause disease, and neutralize them.

The disinfecting agents generally in use rather tend to pollute than to purify the air we breath.

It would be no difficult matter to add an excess of oxygen to the air in the hold of a ship, or any other part of the vessel, by heating peroxide of manganese or chlorate of potass in a suitable apparatus, taking care to have the articles pure, or carbonic acid gas would be disengaged.

The surest way to disinfect a ship and to prevent it from becoming infected, would seem to be to subject the emigrants, by law or voluntary agreement, to the authority of the master of the vessel as fully as soldiers of a regiment are to their commanding officers.

Such heterogeneous materials cannot be reduced to order, united, and made subservient in preventing and removing the causes of disease occurring at sea, unless subject to the same control in coming to this country as they would be after they get here, if called into the field to defend it. A regiment is the type of military organization. Its economy and management applies equally to hospitals and to ships, or to large armies.

If the emigrants were allowed the same wages per diem that the soldiers of continental Europe generally receive, it would better reconcile them to the bridle.

At three cents a day the wages of each during the voyage would seldom exceed a dollar and a half, and the price of the passage could be raised accordingly. It is suggested that, on coming on board, all the adult males should be required to sign a bond agreeing to work at certain regular hours in cleaning and dry rubbing the vessel. This would be the most effectual way to render the air of every part of the vessel dry and healthy. It would likewise be beneficial to the laborers themselves—being removed from the steerage there would be more room for the women and children.

Only that portion of them who were required to perform the manual labor should be entitled to draw wages. But if impracticable or difficult to be enforced, during a state of health, it should nevertheless be insisted upon, if disease made its appearance, as a measure absolutely necessary to disinfect the ship. It should be borne in mind that the fourteenth day after sailing is the most common time for disease to explode on a foul or over-crowded vessel. It generally requires this space of time for preparation of some kind leading to the development of disease.

CANAL STREET, NEW ORLEANS, *February* 27, 1854.

Naval Hygiene, by Samuel A. Cartwright, M. D., New Orleans.

XIV. "A distinction with respect to the number of passengers between vessels passing within the tropics and those not so passing."—Formerly there was a distinction, but the law has been repealed. No sufficient reason is perceived for its re-enactment. The rarefaction of the air produced by heat no doubt led to it, and led to error from not being viewed in connexion with other facts. The truth is, that a ship sailing within the tropics can, with safety, carry more passengers than in the temperate or frigid zones. The reason is, that the cold compels the passengers to shut themselves up when sailing in high latitudes; unless the apartments containing them be capacious, an artificial, unhealthy atmosphere is soon generated. Whereas, within the tropics every room in the vessel is thrown wide open, and this free ventilation prevents the air from being polluted from the breath and perspiration. Ships are going out from this port of New Orleans almost every week so excessively crowded, to Chagres and other places within the tropics, that the passengers could not fail to suffer, both from discomfort and disease, if sailing in a high latitude, cold enough to require them to keep within their several apartments. They have, however, generally a healthy and pleasant voyage, nowithstanding their numbers, because every part of the ship is open, and the most of the passengers, instead of confining themselves to their berths, or to close rooms, as those sailing in a cold climate are compelled to do, come up on the spar-deck to enjoy the refreshing breeze—nearly always stirring in the open sea within the tropics—protected from the rays of the sun and the dews at night by an awning. Many prefer sleeping on deck. Thus more space is always left for those who may choose to remain between decks. The sea-air within the tropics is equally pure as that in extra tropical regions. Near shore it is less unhealthy than coasting in cold climates. The cold land winds (so unhealthy) extend much further into the sea than any winds charged with unhealthy vapors from localities near or within the tropics. When yellow fever is prevailing in New Orleans, boats and ships at anchor before it, or on the opposite shore, less than a mile distant, are but rarely infected with the disease; thus proving that the causes producing such affections are not capable of being transmitted by the winds to any great distance. Many persons unnecessarily suffer in hot climates from tight-fitting garments. The thinner a garment is the hotter it will be if it presses closely to the surface of the body instead of being loose and in folds. The looser and the more folds or plaits a garment has the greater will be the surface for facilitating the cooling process by evaporation. Oil and oleagineous substances so useful in protecting the human system against the cold of the frigid zone when taken internally, are equally useful in protecting it against the heat of the torrid when applied externally. The original natives of hot climates, as Dr. James Johnson has well remarked, have a cutaneous surface of a more oily nature than those of high latitudes. They are also in the habit, which Dr. Johnson says is universal in the east, of making it more so by artificial means, as inunction with various oils—that of the cocoa being preferred. The use of spices and acrid substances as condiments, so far from being injurious, are very useful in preserving health in hot climates. Without them acid drinks and fruits are unhealthy, causing too much perspiration and disordering the bowels. Many customs attributed to ignorance and superstition are founded in the laws of nature. The practice of using highly seasoned food, so common in the torrid zone, is one of them. Mosely has observed that the perspiration in the West Indies is stronger and more acrid than that proceeding from the body of the same individual in England. Hence he infers that a hot climate tends to rob the body of its acrid matter more than a cold one, and requires replenishing by a free use of peppers and other spices. Oat-meal, barley, and such things which figure so conspicuously in the out-fit of ships, and so well relished when made into porridge at the north, become so insipid and disgusting to the appetite in the tropics, as often to cause nausea and vomiting. It is not by re-enacting the repealed law making a distinction between vessels passing within the tropics and those not so passing that health can be preserved, but by ascertaining and observing those irrepealable laws which nature has already made.

NEW ORLEANS, *February* 27, 1854.

No. 21.

Communication from the Mayor of New Orleans.

MAYORALTY OF NEW ORLEANS,
January 25, 1854.

SIR: I have the honor to acknowledge the receipt of your communication of the 29th ultimo, in which, on behalf of a select committee of the United States Senate, appointed to consider the causes and the extent of the sickness and mortality prevailing on board emigrant ships on the voyage to this country, you desire to obtain a list of all the vessels containing emigrant passengers which have arrived at the port of New Orleans during the year 1853, with a designation of the ports whence they sailed, the nationality of the vessel, the number of passengers, the number of deaths, &c. And I have also to acknowledge the receipt of your circular of the same date, making various inquiries with reference to the adequacy of existing laws relative to the transportation of emigrants, and desiring my opinion in regard to the propriety of amending these statutes in some important particulars.

In answer to your first communication, I beg leave to transmit herewith a detailed list of all the vessels that have arrived at this port from abroad with emigrants during the year 1853; giving the date of arrival, the names of the vessels, the port from whence they sailed, the number of passengers, the nationality of the vessels, and the number of deaths that have occurred during the voyage. The result furnishes an aggregate of 37,855 emigrants arrived within the year, the mortality among whom only amounted to 311, or 0.8216 per cent. The proportion, as you will perceive, is very small, and will, no doubt, compare favorably with the other Atlantic cities. It proves, furthermore, that the southern passage is, on the whole, healthier for emigrants than the northern, for the obvious reason that the milder temperature induces the passenger to expose himself much more in the open air, and thus to avoid the consequences of being constantly cooped up in a narrow space and an impure atmosphere, where ventilation at best is considered a matter of secondary importance. The average length of the voyages from the ports specified in the accompanying list was from forty-five to fifty days.

In response to your second communication, I shall proceed to consider the inquiries which you propose separately.

1. The space allotted to each passenger.

In this respect I am of the opinion that the existing laws should be so far changed as to increase the space. There is no doubt in my mind that much of the sickness on board of emigrant vessels is to be ascribed to the crowded manner in which passengers are huddled together, and if the room now allowed were enlarged to the extent of one-fourth, I feel persuaded that the change would be attended with beneficial results.

2. The quantity and the quality of provisions required for each passenger.

In this respect I am not aware of any deficiency in the existing statutes. It is quite true that the change of diet on board of vessels is apt to produce disease among persons who, unaccustomed at home to salt meats, have to satisfy their appetites to a great extent on such food. But I do not well see how this is to be remedied, and the only suggestion I may add is to provide for the strict enforcement of the law, so as to afford to the emigrant the requisite amount of provisions which he is entitled to.

3. The permission allowed to the passengers to furnish their own provisions for the voyage, instead of making it, in all cases, the duty of the master to provide them.

I am decidedly of opinion that it should be made the duty of the captain to furnish the requisite provisions in all cases. By allowing the passengers to provide for their own wants, they frequently embark utterly destitute of all means, and with scarcely a sufficiency of provisions to last half the voyage. The stores which they lay in, moreover, are often bad and tainted, and of necessity are calculated to engender disease; whereas if, in every instance, the master was compelled to give the required amount of good and wholesome food, one material cause of sickness would be done away with.

4. Ventilation.

5. The cooking arrangements.

In both of these respects I should think that the existing laws, if adequately enforced, would answer all purposes.

6. The duty of the master to enforce personal cleanliness and to insure the cleanliness of the vessel.

The captain should, in every instance, be clothed with the most ample authority to enforce cleanliness. It is notorious that a large portion of the emigrants are not of the most cleanly habits, and that if allowed to remain undisturbed, they will permit dirt and filth to accumulate in the hold to an extent dangerous to the health of all the inmates. The only remedy is to enforce cleanliness and to grant plenary power to shipmasters to carry it into effect.

There can be no question in regard to the propriety of amending the existing laws in conformity with the requirements and suggestions contained in your circular of the 29th December. The employment of a qualified and experienced surgeon would in itself be a vast improvement, and would certainly be the means of saving many lives. So, also, the limitation to the number of passengers agreeably to the opinion which I expressed in answer to inquiry No. 1.

In addition to this I believe that when disease breaks out on board of an emigrant vessel, provision should be made to separate the sick and to keep them as much apart as possible. There can be no doubt that the constant contact between the indisposed and those in a healthy condition, even when the disease is not of a contagious character, is prejudicial, and any means by which a seclusion can be brought about, would prove to be an amelioration.

During the course of my experience, I have ascertained that much of the sickness on board of emigrant vessels was due to the reckless and inhuman manner in which passengers are not unfrequently shipped from the other side. It is too often the case that emigrants are taken on board already sick and enfeebled, or partially indisposed, and the change of diet and air frequently brings on disease in a worse shape, terminating in death. Their own lives are thus made to pay the forfeit of this gross neglect, whilst those of their fellow-passengers become endangered from contact, and thus we see vessels severely afflicted with disease and death, which, but for the grasping conduct of shipping brokers abroad, would have been kept entirely free from sickness.

I have endeavored to lay before you all the information in my power in response to your inquiries, and I shall be most happy to furnish you, at all times, with such further facts as you may desire.

I have the honor to be, with great respect, your obedient servant,

G. D. CROSSMAN, *Mayor.*

Hon. Hamilton Fish,
Chairman, U. S. Senate, Washington.

www.ingramcontent.com/pod-product-compliance
Lightning Source LLC
LaVergne TN
LVHW021409110826
845150LV00007B/1851